ISBN 978-0-260-60929-8
PIBN 11119578

1 MONTH OF
FREE
READING

at
www.ForgottenBooks.com

By purchasing this book you are eligible for one month membership to ForgottenBooks.com, giving you unlimited access to our entire collection of over 1,000,000 titles via our web site and mobile apps.

To claim your free month visit:
www.forgottenbooks.com/free1119578

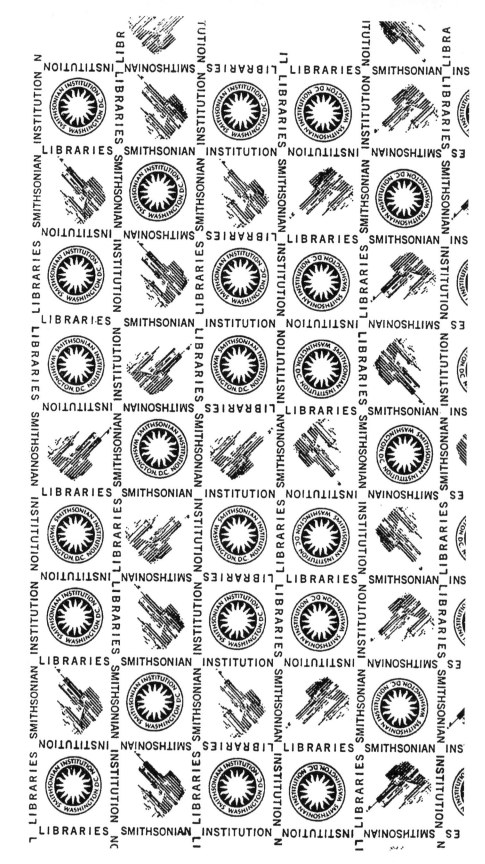

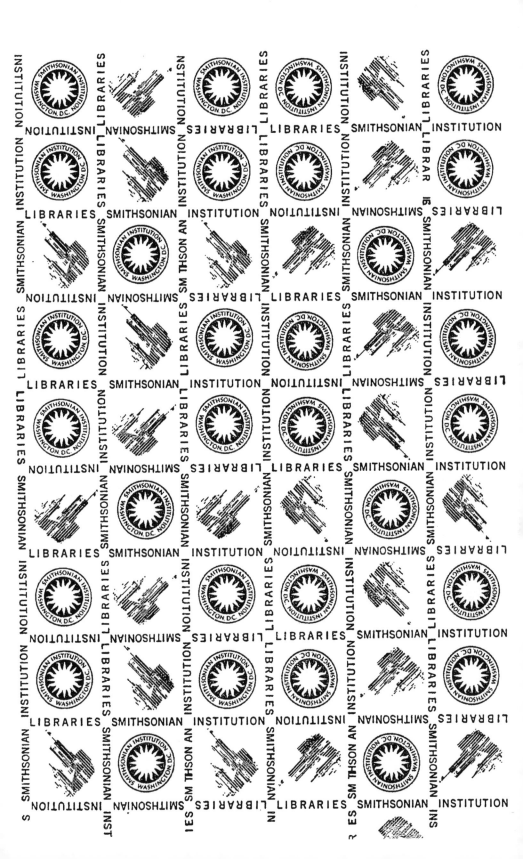

FOURTH
SUPPLEMENT
1933 to 1938 (inclusive)

to

THE LENG CATALOGUE

of

COLEOPTERA

of

AMERICA, NORTH OF MEXICO

By
RICHARD E. BLACKWELDER
Assistant Curator
in charge of
Coleoptera
AMERICAN MUSEUM OF NATURAL HISTORY

―――――

Mount Vernon, N. Y.
JOHN D. SHERMAN, JR.
December
1939

FOURTH
SUPPLEMENT
1933 to 1938 (inclusive)

to

THE LENG CATALOGUE

of

COLEOPTERA

of

AMERICA, NORTH OF MEXICO /

By
RICHARD E. BLACKWELDER
Assistant Curator
in charge of
Coleoptera
AMERICAN MUSEUM OF NATURAL HISTORY

———

Mount Vernon, N. Y.
JOHN D. SHERMAN, JR.
December
1939

Buchanan

Printed in the United States of America
by
THE FREYBOURG PRINTING CO.
Mount Vernon, N. Y.

INTRODUCTION

This fourth supplement to Mr. Leng's catalog of North American Coleoptera is offered by the compiler and by the publisher with the hope that it may help to alleviate a condition which has led to a widespread desire for a complete new catalog. Since there appears to be little chance of financing such a new catalog in the near future, it has been thought worth while to prepare another supplement which would fill some of the more obvious needs.

It is probable that a great many entomologists have no conception of the amount of work involved in the preparation of a bibliographic catalog for publication. The writer, when he undertook the present task, anticipated at most a two or three month job. Work was started about the first of March and has occupied most of the compiler's time continuously till the middle of October. It is doubtful if the work could have been undertaken if it had been known that it would involve such an outlay in time and study.

Although this supplement includes many changes in make-up from that followed in the previous supplements (as will be explained below), it has been the endeavor of the compiler to make it as useful as possible in connection with the original catalog and the other supplements. Although Mr. Leng has not had an active part in this work, all proposed changes in procedure have been examined by him and do, I believe, have his complete approval.

In my opinion it is due principally to the excellence of the original catalog that it is possible to issue supplements which can give any satisfaction. Few coleopterists have contributed so largely to the work of all their colleagues as Mr. Leng, and few large works have been so completely constructive in their nature or so universally indispensable as the "Leng Catalog".

It would seem to be axiomatic that the fourth of a series of supplements should conform in procedure and treatment with the previous parts. In spite of this a glance at the present work will suffice to show that it departs widely from the other supplements. The reasons for these changes are outlined below.

The primary cause of changes in make-up of the supplement was the desire to save space. The complete division of the page into two columns and the omission of numbers have together increased the space available by twenty to thirty percent. The first of these changes will scarcely be noticed by users of the catalog. The second may meet with disapproval of some, but in its defense may be mentioned the substantial amount of space saved which has been used for the inclusion of revisionary studies.

The second important departure from the previous supplements is in the matter of authorities for new synonyms, records, and other changes. It has been felt that a summary of the compiler's opinions on the status of species and classifications (such as the original catalog properly included to some extent) was of less interest in the supplement than an accurate reporting of what has been published during the last six years and who published it. Accordingly nearly every new proposal is accompanied by a citation to the author of that proposal.

It has been the aim of the compiler to include all new species described up to January, 1939, all synonymy proposed up to the time the manuscript was completed (about September 1, 1939), all new records for North America with synonyms, all contributions to the classification of beetles, and all revisionary studies, whether of families, genera, or groups of species.

Bibliographic references are made as in the catalog by giving the author, date, and page. An indication of distribution is given. Italics have been used always for synonyms, and generic synonyms are in addition placed in parenthesis. Names which were proposed for categories less than species are preceded by a letter denoting the category assigned to them by the writer quoted. These are arranged in indented form to emphasize their status lower than species.

For the first time subgenera are definitely indicated as such. These names are strongly indented and are printed in Roman capitals. Space considerations prevented the insertion of the word subgenus or an abbreviation in these cases, but once the system of type faces is understood, there should be no difficulty. Bold-face is used always for genera, italics for synonyms, and capitals for subgenera. (In the Carabidæ only there have recently been proposed some names for "sections" within certain subgenera. These have been printed in small capitals and still more indented than the subgenera.)

Two uses are made of the asterisk (*). Wherever there has been recently published a presumably complete revision of a genus, an asterisk is placed at the left hand margin immediately below the generic name. Any other items which were not included in the revision are separated from the latter by a row of three asterisks. Thus the three asterisks always denote the end of a revision or of a subgenus, or of a special rearrangement of genera as mentioned in a footnote.

One of the characteristics of human endeavor is the occasional mistake that is almost inevitably to be found. The most concentrated effort will not prevent a few errors from marring any work. These are generally taken for granted and are scarcely noticed unless they are present in unusual numbers. While examining the sources of the present compilation, the writer was forced to the conclusion that an entirely disproportionate number of the errors in our taxonomic papers on beetles has been made by a relatively small number of workers. For the most part these are errors due to carelessness or indifference, and they usually result from failure to ascertain all the facts or from one's being in too great a hurry to get papers published. A few of these are mentioned in footnotes in the text. Among the more common and very unfortunate errors may be mentioned the following: The renaming of preoccupied genera and species for which there was already one or more available names; failure to give the author of a homonym an opportunity to rename it; changing the name of a genus or species because of page precedence; citing a synonym in parenthesis between the generic and specific names, as *Mylabris (Bruchus) atomus*, where *Bruchus* is not used as a subgenus; failure to follow the Rules strictly in the matter of accepting or correcting the original spelling of names; and extreme carelessness in giving references to older species and in the spelling of the names themselves.

Most of these errors have been copied into the supplement without specific comment, but attention is directed to some by footnotes. In most cases in which there have been changes made by the compiler, this fact is indicated. But in general responsibility and authority for all items are placed directly back on the entomologist who published them. An exception to this statement must be noted in the matter of the bibliographic references. Many of these have been corrected by independent study, and wherever there was a discrepancy the original catalog has been followed (unless ample reasons have been published for the change).

Many of the entomologists whose recent work is included in this supplement were kind enough to check the preliminary lists of their publications. One hundred and one such lists were sent out, and about 85 were returned with the necessary additions or corrections. Many important oversights were thus corrected, and many new items were contributed by these workers from their own studies. Grateful acknowledgment is made to all these friends.

SIGNS AND ABBREVIATIONS

* indicates a revision of the genus (or other group).
*** indicates the end of a revision or of a subgenus.
[before locality indicates that it applies to line above.

a. aberration.
n. natio, used by Breuning as a category between subspecies and aberration.
s. subspecies.
v. variety.

FOURTH SUPPLEMENT
TO CATALOGUE OF COLEOPTERA
OF AMERICA, NORTH OF MEXICO

ADDITIONS AND CORRECTIONS TO

DECEMBER 31, 1938

CICINDELIDÆ

Omus

humeroplanatus W. Horn 10-263
 Loc. should be Del Norte Co., Cal.[1]
submetallicus G.Horn 68-129
 s.niger Cazier 37-94   Cal.

Megacephala Latr. 02-79

              TETRACHA Hope 38-6 [2]
affinis Dej. 25-12
 s.angustata Chevr. 41-55   C.A. Tex.[2]
virginica Linn. 66-657
 v.melæna Cartw. 35-70   S.C. Va.

Cicindela

purpurea Cliv. 90-14   E.U.S.-Sierras
 marginalis Fahr. 01-240 [3]
 spreta Lec. 57-37 (not Lec. 1848) [3]
 s.auduboni Lec. 45-207 (in part)[3]
                        Ill.-Ut.-Alb.-Minn.
 graminea Schp. 83-89 (in part)[3]
 v.lauta Csy. 97-296 [3]   Ore. Cal.
  franciscana Csy. 13-23 [3]
  graminea Schp. 83-89 (in part)[3]
 v.mirabilis Csy. 14-358 [3][4]
 v.nigerrima Leng 18-139 [3]
                        Mass.-N.M.-Mont.
  auduboni Lec. 45-207 (in part)[3]
 cimarrona Lec. 68-49 [3]
                   N.M. Colo. Ariz. Ut.
 v.sedalia Smyth 33-201 [3][4] Colo. (Ore.)

---

[1] Nunenmacher in litt.
[2] Darlington—85.
[3] Nicolay in litt.
[4] Nicolay—84.

limbalis Klug 34–39 [3]   Mass.-Ky.-Kan.-
 amoena Lec. 48-177 [3]   [Alb.
 splendida || Lec. 57-36 [3]
 militaris Varas 28-242 [3]
 s.limbalis (s.str.)[3]
  v.spreta Lec. 48-177 [3]
  v.transversa Leng 02-131 [3]
                    Kan. Mo. D.C. N.J.
 s.awemeana Csy. 13-23 [3]   Man.
  v.eldorensis Csy. 13-23 [3]   Ore.
 s.auguralis Csy. 13-21 [3]   Colo. N.M.
  inducta Csy. 13-22 [3]
  ardelio Csy. 13-21 [3]
splendida Hentz 30-254 [3][4]   N.C.-Ark.-
 discus Klug 34-23 [3]   [Colo.-Neb.
 s.splendida (s.str.)
  v.cyanocephala Varas 29-239 [3]
                             Neb. Kan.
  amoena of Harris 11-8 [3]
 s.denverensis Csy. 97-297 [3] Colo. Neb.
  graminea of Csy. 13-21 [3] [Kan.
  v.conquisita Csy. 14-357 [3][4] Neb.Colo.
   oreada Csy. 14-358 [3][4]
   plattensis Smyth 33-202 [3][4]
  v.pugetana Csy. 14-20 [3]
                     B.C.-Wash.-Mont.-Wyo.
  v.propinqua Knaus 22-194 [3] Nev.Cal.
   arida A. C. Davis 28-65 [3]
  v.ludoviciana Leng 02-131 [3] Ark. La.
 decemnotata Say 17-19 [3]   Alas.-Cal.-
  lantzi Harris 13-68 [3]   [Alb.-Kan.
  albertina Csy. 13-24 [3]
                    • • •
 plutonica Csy. 97-296 [3]
  s.leachi Cazier 36-124   Cal.
 senilis Horn 66-395   Cal.
  exoleta Csy. 09-272 [5]
  frosti Varas 27-174 [5]

---

[5] Cazier—37.

willistoni Lec. 79-507   Wyo.
s.echo Csy. 97-298 [5]   Ut. Cal. Nev.
amadensis Csy. 09-272 [5]
s.pseudosenilis W. Horn 00-117 [6]
    Cal. Nev.
s.spaldingi Csy. 24-14 [5]   Ut.
fulgida Say 23-141   Neb.-N.M.-Wyo.
wallisi Calder 22-191 [5]
azurea ¦ Calder 22-62 [5]
s.westbournei Calder 22-62 [5]   Man.
elegans ¦¦ Calder 22-62 [5]
v.subnitens Calder 22-62 [5]   Neb.
s.pseudo-willistoni W.Horn 38-13 [6] Wy.
parowana Wickh. 05-165   Ut. Wash.
remittens Csy. 24-14 [5]
platti Cazier 37-161 [3][7]   Cal.
sexguttata Fahr. 75-226
s.denikei Brown 34-22   Ont.
politula Lec. 75-159
s.lætipennis W.Horn 14-32 [8] Mex. Tex.
viridisticta Bates 81-14
s.arizonensis Bates 84-260 [9] Ariz. Mex.
cursitans Lec. 57-60
alata Lilj. 32-215 [9]
pusilla Say 17-21
s.lunalonga Schp. 83-122 [5]
v.wagneri Cazier 37-117 [5]   Cal.
californica Men. 44-52
s.mojavi Cazier 37-116   Cal.

CARABIDÆ

Trachypachus Mots. 45-86
(*Trachypachys* G. & H. 68-46)[10]
(*Trachypachis* Lac. 54-47)[10]
zetterstedti (Gyll.) 27-417   Holarctic
transversicollis Mots. 44-86 [10]
laticollis Mots. 64-194 [10]
inermis Mots. 64-194 [10][11]
holmbergi Mannh. 53-119 [10]
oregonus Csy. 20-145 [10]
specularis Csy. 20-146 [10]

• • •

CYCHRINI [12]

Scaphinotus Latr. 22-87 [3]
IRICHROA Newm. 38-385
(*Megaliridia* Csy. 20-175)[13]
andrewsii Harris 39-195
s.andrewsii Harr (s.str.)[12] N.C. Tenn.
s.æneicollis Beut. 03-513 [12] N.C. Tenn.
purpurata Beut. 18-89 [12]
v.tricarinata Csy. 14-25 [13]
s.violacea Lec. 63-4 [12] Ga. N.C.
amplicollis Csy. 20-174 [12]
s.germari Chd. 61-495 [12] Pa. Va. Tenn.
mutabilis Csy. 20-173 [12]
longicollis Csy. 20-173 [12]
modulata Csy. 20-174 [12]

guyoti Lec. 66-363   N.C.
v.angelli Beut. 18-89 [12]
ridingsi Bland. 63-353
s.ridingsi (s.str.)[12]   Va. D.C.
s.monongahelæ Leng. 17-36 [12] Pa. Ten.
tenuiceps Csy. 20-173 [12]
vidua Dej. 26-12
s.leonardi Harris 39-193 [12]   N.Allegh.
s.vidua (s.str.)[12]   Allegh.
unicolor Knoch 01-187 [12]
s.irregularis Beut. 03-513 [12]   N.C.
SCAPHINOTUS (s.str.)[12]
snowi Lec. 81-74   N.M.
rœschkei VanD. 07-118   Ariz.
fuchsi Rœsch. 07-570   Ariz.
vandykei Rœsch. 07-136   Ariz.
corvus Fall 10-89   Ariz.
petersi Rœsch. 07-118   Ariz.
mexicanus Bates 82-320   Ariz. Mex.
kelloggi Dury 12-104   N.M.
biedermanni Rœsch. 07-571   Ariz.
elevatus Fahr. 87-198
s.elevatus (s.str.)[12]   E.U.S.
s.tenebricosus Rœsch. 07-119 [12] E.U.S.
unicolor || Lec. 53-398 [12]
heros Lec. 48-440 [12]
s.flammeus Hald. 44-54 [12]   S.U.S.
dilatatus Lec. 53-398 [12]   [C.U.S.
s.floridanus Leng 15-564 [12]   Fla.
coloradensis VanD. 07-141   Colo.
unicolor Fahr. 87-198
s.unicolor (s.str.)[12]   C.U.S.
s.heros Harris 39-196 [12]   O. Ind.
s.shoemakeri Leng 14-143 [12]   D.C.

Brennus Mots. 65-311 [12]
• BRENNUS (s.str.)
cordatus Lec. 53-399   Cal.
v.vernicatus Csy. 20-183 [12]
v.rufitarsis Csy. 20-184 [12]
marginatus Fisch. 22-79
s.marginatus (s.str.)[12]
    Alas. B.C. Wash.
v.gracilis Gehin 85-76 [12]
v.insularis Csy. 97-334 [12]
v.cupripennis Csy. 97-334 [13]
v.confusus Csy. 97-336 [12]
v.wrangeli Csy. 20-182 [12]
s.fulleri Horn 78-173 [12]   Ore.
v.fallax Rœsch. 07-174 [12]
v.montanicus Csy. 20-182 [12]
s.columbianus Csy. 20-180 [12]   B.C.
s.oregonus Csy. 20-182 [12]   Ore.
interruptus Men. 44-54
s.interruptus (s. str.)[12]   Cal.
sinuatus Csy. 97-330 [12]
s.compositus Csy. 97-332 [12]   Cal.
s.constrictus Lec. 53-398 [12]   Cal.
interruptus Lec. 68-60 [12]
dissolutus Csy. 97-329 [12]
corpulentus Csy. 97-331 [12]
parvulicollis Csy. 20-176 [12]
s.dissolutus Schaum 63-72 [12]   Cal.
porcatus Csy. 97-328 [12]
s.politus Csy. 97-330 [12]   Cal.
s.montereyensis Csy. 20-177 [12]   Cal.
s.procerus Csy. 20-179 [12]   Cal.

[3] Nicolay in litt.
[5] Cazier—37.
[6] W. Horn—38.
[7] Cited as subspecies by Cazier—37.
[8] Benedict—34.
[9] W. Horn—35.
[10] Hatch—33.
[11] Listed as valid by Csiki & Hetschko—33.

[12] Revision of tribe. Vacher de Lapouge—32. Alternate arrangement of certain genera by Van Dyke and Valentine below.

[13] See alternate arrangement by Lapouge, above.
[14] Revision of subgenus, Van Dyke—38.
[15] Arrangement of subgenera and species, Valentine—35.
[16] Leng Catalog.
[17] Revision of subspecies, Valentine—36.

s.darlingtoni Val. 35-356 [15] [17]   Ten. N.C.
s.barksdalei Val. 36-230 [17]   N.C.
s.germari Chd. 61-495 [15] [17]
              Pa.-N.C.-Ky.
s.mutabilis Csy. 20-173 [17]   Pa. W.Va.
   longicollis Csy. 20-173 [17]   [O. Ky.
   modulata Csy. 20-174 [17]
s.waldensia Val. 35-357 [15] [17]   Tenn.
violacea Lec. 63-4 [15]
   s.violacea (s.str.) [15]   Ga. N.C.
   s.carolinæ Val. 35-358 [15]   N.C.
   æneicollis Beut. 03-513 [15]   N.C.
   tricarinata Csy. 14-25 [15]   N.C. Tenn.
   guyoti Lec. 66-363 [15]   N.C. Va.
   v.angelli Beut. 18-89 [15]   N.C.
   confusus Dari. 31-146 [15]
lodingi Val. 35-364
   s.lodingi (s.str.) [15]   Ala.
   s.obscura Val. 35-366 [16]   Ala.
ridingsi Bland 63-354
   s.ridingsi (s.str.) [15]   Va.
   s.monongahelæ Leng. 17-36 [15]   Pa.
   tenuiceps Csy. 20-172 [15]
   s.intermedia Val. 35-368 [15]   Va.

* * *

* NOMARETUS Lec. 53-399 [18] [19]
bilobus Say 25-73 N.H. N.Y. Can. Mich.
liebecki VanD. 36-40   Tex.
fissicollis Lec. 53-399 Ill. Kan. Mo. Tex.
cavicollis Lec. 59-3 Kan.Ark.Mo.Tx.La.

* * *

Pseudonomaretus Rœschke 07-121 [19]
*   (*Maronetus* Csy. 14-30) [19]
merkeli Horn 90-71 [20]   Ida.
   idahoensis Webb. 01-133 [20]
relictus Horn 81-188 [20]   Ida. Wash.
   v.regularis Lec. 85-2 [20]
imperfectus Horn 60-569 [20]   N.C. Pa.
hubbardi Schw. 95-272 [20]   N.C.
incompletus Schw. 95-271 [20]   Va.
debilis Lec. 53-399 [20]   Ga. Car.
   v.alpinus Beut. 03-512 [20]   N.C.
schwarzi Beut. 13-139 [21]   N.C.

Neocychrus
longiceps VanD. 24-5 (Brennus) [22]

Sphæroderus [19]
canadensis Chd. 61-498
   s.canadensis (s.str.) [22]   Ont. H.B.T.
   palpalis Mots. 65-312 [23]   [N.Eng.
   blanchardi Leng. 16-41 [23]   [N.C.
   s.lengi Dari. 33-63 [23]   N.C. Tenn.

---

[15] Arrangement of subgenera and species, Valentine—35.
[17] Revision of subspecies, Valentine—36.
[18] Revision of subgenus, Van Dyke—36.
[19] Listed as valid genus by Lapouge—32, with Sphæroderus (part) as synonym and subgenus Pseudonomaretus Rœschke (Maronetus Csy.=).
[20] Vacher de Lapouge—32. (Revision of part of tribe with arrangement differing from that of Breuning.)
[21] Listed as synonym by Nicolay & Weiss—34.
[22] Van Dyke in litt.
[23] Darlington—33.

GEHRINGIINI

Gehringia Darl. 33-110
* olympica Darl. 33-111   Wash. Mont.

CARABINI

Carabina [24]

Carabus Linn. 58-413
* (*hummeli* Fisch., not N. American) [24]

   CARABUS (s.str.) [24]
     Eucarabus Geh. 85-xxi [24]

chamissonis Fisch. 20-88 Alas.-Greenl.-
   brachyderus Wiedem.24-110 [24] [N.H.
   groenlandicus Dej. 31-554 [24]
   washingtoni Csy. 20-155 [24]
vinctus Web. 01-42 [25] N.Y.-Ga.-Ind.-Ont.

   ligatus Germ. 24-6 [24]
   interruptus Say 25-76 [24]
   carinatus Dej. 26-79 [24] [26]
   georgiae Csiki 27-185 [24]

limbatus Say 25-77 [26] N.Y.-Ga.-Ill.-Ont.
   goryi Dej. 31-544 [24]

mæander Fisch. 22-103 [27]   Sib. N.A.
   incompletus Fisch. 27-303 [24]
   ehrenbergi Fisch. 29-368 [24]
   palustris Lec. 48-444 [24]
   hudsonicus Mots. 65-293 [24] [26]
   lapilayi || Dohrn 78-362 [24]
   simoni Heyd. 79-163 [24] [26]
   excatenatus Kr. 80-337 [24]
   excostatus Kr. 80-338 [24]
   obscuratus Geh. 85-26 [24]
   lecontei Geh. 85-27 [24]
   mongolicus Lap. 05-305 [24]
   lecontei Angeli 14-75 [24]

   n.lapilayi Cast. 35-89 [24] [26]   Newf. St.P.
   atlanticus Lap. 24-191 [24]   [& M.
     Archicarabus Seidl. 87-6 [24] [29]

nemoralis Müll. 64-21 [30]
   n.nemoralis (s.str.) [24]
   foetens Voet 78-71 [24]
   hortensis || Fahr. 75-237 [24]
   nigrescens Letz. 50-85 [24]
   virescens Letz. 50-85 [24]
   tristis DallaT. 77-27 [24]
   krasae Roub. 03-380 [24]
   brunnipes Lap. 08-19 [24]
   deletus Lap. 08-20 [24]
   canadensis Lap. 08-19 [24] [26]
   auratus Heuer 26-332 [24]
   lestagei Basil 30-62 [24]

---

[24] Revision of subtribe, Breuning—32-37.
[25] Placed by Lapouge—32 in genus Lichnocarabus Reitt. 96-161. (Synonyms Mesocarabus Gehin 85-16 and Limnocarabus Gehin 85-27).
[26] Listed as subspecies by Lapouge—32.
[27] Placed in subgenus Paracarabus Lap. 32-530 by Lapouge—32.
[29] Listed as genus by Lapouge—32.
[30] Vacher de Lapouge—32 (Revision of part of tribe with arrangement differing from that of Breuning.

OREOCARABUS Geh. 76-12 [24]
 OREOCARABUS (s.str.)[24]
tædatus Fahr. 87-196　Alas.-N.M.-Man.
s.tædatus (s.str.)[24]
seriatus Wied. 21-109 [24]
baccivorus Fisch. 23-87 [24]
gladiator Mots. 65-285 [24]
patulicollis Csy. 13-57 [24]
franciscanus Csy. 13-58 [24]
bicanaliceps Csy. 20-154 [24]
stocktonensis Csy. 20-155 [24]

s.agassii Lec.50-209 [24]　Ore.-N.M.-Man.
oregonensis Lec. 54-16 [24]
gladiator || Heyd. 79-161 [24]
canadicus Rœsch. 00-68 [24]
montanicus Csy. 13-58 [24]
s.coloradensis Breun. 33-719 [24]　Colo.

DIOCARABUS Reitt. 96-185 [32]
lutshnikianus Basil. 37-63　Alaska
ORINOCARABUS Kr. 78-328 [24]
truncaticollis Esch. 29-22 [33]　Alas., Sib.
s.alaskensis Basil. 37-63　Alaska

MESOCARABUS Thoms. 75-640 [24]
problematicus Hbst. 86-177
s.gallicus Geh. 85-15 [24]　Eur.Asia.Cal.
n.gallicus (s.str.)[24]
intricatus || Cliv. 95-20 [24]
austriacus Strm. 15-78 [24]
dissitus Fisch. 27-170 [24]
californicus Mots. 65-287 [24]
beauvoisi Kr. 78-158 [24]
catenulatus of authors [24]
problematicus Lap. 16-81 [24]
scandinavicus Born. 26-65 [24]
cruris Csiki 27-79 [24]
lhommei Hoffm. 28-12 [24]

HEMICARABUS Geh. 85-xix [24][30][34]
HEMICARABUS (s.str.)[24]
serratus Say 25-77 [30]　Que.-Va.-N.M.-B.C.
lineatopunctatus Dej. 26-77 [24]
ligatus Kby. 37-18 [24]
canadensis Meish. 53-10 [24]
vegasensis Csy. 13-59 [24]
s.tatumi Mots. 65-293 [24]　H.B.T. Newf.

EURYCARABUS Geh. 85-xxi [24]
TANAOCARABUS R. 96-135 [24]
forreri Bates 82-320 [35]　Mex. Ariz.
sylvosus Say 25-75　Vt.-Fla.-Kan.
lherminieri Dej. 26-151 [24]
finitimus Hald. 52-373　Ut.Tex.Okla.Ala.
lecontei Csy. 13-57 [24]
caseyi Ang. 14-75 [24]

PRCCRUSTES Bon. 09-39 [24]
MEGODONTES Sol. 48-58 [24]
vietinghoffi Ad. 12-170
s.vietinghoffi (s.str.)[24]
n.vietinghoffi (s.str.)[24]　Sib. Alas.
[B.C. H.B.T.
(n.fulgidus Fisch., not N. Amer.)[24]

[32] Basilewsky—37.
[33] Placed in Diocarabus by Basilewsky—37.
[34] Reference given as 76-25 by Gehin.
[35] Van Dyke—38.

Calosoma
* CALLITROPA Mots. 65-300 [34]
protractum Lec. 62-52
dolens Chd. 69-376 [36]

* * *

morrisoni Horn 85-128　Colo.
prominens Lec. 53-400
v.parvicolle Fall 10-90 [36]
clemens Csy. 14-32 [36]
subgracile Csy. 13-63 [36]
pertinax Csy. 20-163 [36]
glabratum Dej. 31-565　S.Am. C.Am.
v.sponsum Csy. 97-340 [36]
eremicola Fall 10-91 [36]
incertum Lapouge 24-38 [36]
hospes Csy. 13-63 [36]
parviceps Csy. 97-341 [36]
rugosipenne Schffr. 11-113 [36]

PARACALOSOMA Br. 27-145 [36]
palmeri Horn 76-199 [36]

* * *

affine Chd. 43-746
v.triste Lec. 44-201 [36]
v.tristoides Fall 10-92 [36]
calida Fab. 75-237
v.stellata Csy. 97-344 [36]
concreta Csy. 20-157 [36]
tepidum Lec. 49-199
pallax Csy. 20-160 [36]
cancellatum Esch. 29-23　Cal.
præstans Csy. 20-159 [36]
rectilatera Csy. 20-158 [36]
sagax Csy. 20-158 [36]
aenescens Lec. 54-16 [37]
moniliatum Lec. 49-200
bicolor Walk. 66-313 (Carabus)[36]
vancouverensis Csiki 27-286
[(Carabus) [36]

Callisthenes
discors Lec. 57-31
v.dietzi Schffr. 04-197 [36]
v.schæfferi Breun. 28-80 [36]
irregularis || Schffr. 15-235 [36]
striatus Breun. 28-86
striatulus || Lec. 59-4 [36]
luxatus Say 23-149
klamathensis Csy. 20-169 [36]
v.zimmermanni Lec. 49-445 [36]
parowanus Csy. 20-167 [36]
debilis Csy. 20-167 [36]
tegulatus Csy. 13-72 [36]
viator Csy. 13-72 [36]
v.pimelioides Walk. 66-312 [36]
pustulosus Csy. 13-73 [36]
a.striatus Breun. 28-86 [36]
reflexus Csy. 20-164 [36]
semotus Csy. 20-166 [36]
utensis Csy. 20-165 [36]
a.subasperatus Schffr. 15-235 [36]

[36] Csiki and Hetschko—33.
[37] Leng Catalog.

Elaphrus Fahr. 75-227
    TRICHELAPHRUS Sem. 26-39
horni Csiki 27-420 [36]   Cal.
   viridis Horn 78-52 [36]

Nebria Latr. 06-221
    NEBRIA (s.str.) [37]
crassicornis VanD. 25-121
meanyi VanD. 25-118
piperi VanD. 25-117
schwarzi VanD. 25-116
    NEBRIOLA Dan. 03-164 [36]
lyelli VanD. 25-120
riversi VanD. 25-115
spatulata VanD. 25-119
kincaidi Schw. 00-525 [36]
  columbiana Csy. 13-48 [36] [38]
paradisi Darl. 31-24 [38]
  kincaidi | Bänn. 25-264 [36]
  vandykei || Dari. 30-104 [38]

Oxydrepanus Putz. 66-103
* rufus Putz. 46-564 [39]   Cuba, Fla.
  brevicarinatus Putz. 46-571 [40]

Clivina
oregona Fall 22-164 (Dyschirius) [41]

Goniotropis Gray 32-273 [42]
* parca Lec. 84-2 [43]   Ariz.

Psydrus Lec.
*
    (*Monillipatrobus* Hatch 33-118) [44]
piceus Lec. 46-154   Cal.-L.Sup.
  punctatus Hatch 33-118 [44]

Bembidion
    PLATAPHUS Mots. 64-184 [45]
planiusculum Mannh. 43-215
  basicorne Notm. 20-185 (Microme-
          [lomalus) [43]
nigrum Say 23-85 [45]
quadrulum Lec. 61-340 [45]
recticolle Lec. 63-19 [45]
concolor Kby. 37-54 [45]
    EUPETEDROMUS Net. 11-190 [45]
lengi Notm. 19-98 [45]
    PLATAPHODES Gglb. 92-152 [45]
kuprianovi Mannh. 43-217
  bucolicum Csy. 18-34 (Trechone-
          [phra) [45]

• • •

planum Hald. 43-303
  vulsum Csy. 18-55 [46]
  filicorne Csy. 18-56 [46]

TRICHOPLATAPHUS Net. 14-
              [51 [45]
fugax Lec. 48-467 [45]
  champlaini Csy. 18-56 [46]

• • •

grapei Gyll. 27-403 [47]   Grecni.
caducum Csy. 18-80
  albidipenne Csy. 18-80 [46]
  petulans Csy. 18-81 [46]
imperitum Csy. 18-81
  prociduum Csy. 18-91 [46]
  ?simulator Csy. 18-93 [46]
    FURCACAMPA Net. 31-158 [45]
affine Say 23-86 (Notaphus) [45]

• • •

anguliferum Lec. 52-185   N.S.-N.Eng.-
  habile Csy. 18-162 [46]   [B.C.-Cal.

ANILLINI

Anillina [48]

Anillinus Csy. 18-167
*
  (*Anillaspis* Csy. 18-168) [48]
explanatus Horn 88-26   Cal.
debilis Lec. 53-397   Cal. Tex.
fortis Horn 68-127   D.C.-Ga.-Tenn.
  carolinae Csy. 18-168 [48]
dobrni Ehlers 84-36   Fla.
affabilis Brues 02-361   Tex.

• • •

Tachys

nanus Gyll. 10-39
  inornatus Say 23-88 [45]
  rivularis Mots. 50-8 [45]

PATROBINI [49]

Patroboidea VanD. 25-67
* rufa VanD. 25-69   Alas.-Wash.

Platypatrobus Darl. 38-146
* lacustris Darl. 38-146   L.Sup.

Diplous Mots. 50-10
*
  (*Platidius* Chd. 71-51) [50]
californicus Mots. 59-123   Cal.-Mont.-
  trochanterinus Lec. 69-375 [49] [B.C.
  latipennnis Csy. 18-399 [49]
  incisus Csy. 18-399 [49]
  strenuus Csy. 18-400 [49]
  rectus Csy. 18-400 [49]
  sierranus Csy. 18-401 [49]
  breviusculus Csy. 18-401 [49]
aterrimus Dej. 28-32   Alas.-Ore.-Colo.
  fulcratus Dej. 69-374 [49]
  breviceps Csy. 18-402 [49]
  tenuitarsis Csy. 18-403 [49]
  coloradensis Csy. 18-403 [49]
  reflexus Csy. 18-403 [49]

---

[36] Csiki and Hetschko —33.
[37] Leng Catalog.
[38] Bänninger—33.
[39] Darlington—35.
[40] Csiki—Col. Cat. p. 92.
[41] Fall in litt.
[42] Listed as a valid genus by Csiki & Hetschko—33, but placed in tribe Ozænini rather than Pana-gæiini as by Leng.
[43] Csiki & Hetschko—33.
[44] Hatch—35.
[45] Csiki—Col. Cat. p. 126.
[46] Nicolay & Weiss—34.

[47] Cited as valid by Henriksen—35.
[48] Revision of subtribe. Jeannel—37.
[49] Revision of tribe, Darlington—38.
[50] Csiki—Col. Cat. p. 98.

filicornis Csy. 18-404   Cal.-B.C.
  aterrimus of Horn (part)[49]
rugicollis Rand. 38-560   N.E.N.A.
  angicollis Rand. 38-1 (missprint)[49]
  longipalpis Notm. 19-231 [49]

Patrobus Dej. 21-10
*   NEOPATROBUS Darl. 34-155
longicornis Say 25-40   Newf.-B.C.-Fla.-
  americanus Dej. 28-34 [49]   [Ariz.
lecontei Chd. 71-47
  s.lecontei (s.str.)[49]   Colo.Alb.L.Sup.
    rufipes || Lec. 63-18 [49]   [H.B.T.
    septentrionis Horn 75-130 (part)[49]
    canadensis Csy. 24-67 [49]
  s.gravidus Dari. 38-159 [49]   Newf.
fossifrons Esch. 23-104
  s.fossifrons (s.str.)[49]   Alaska
    longiventris Mannh. 53-145 [49]
    fulvus Mannh. 53-145 [49]
    latiusculus Chd. 71-46 [49]
    septentrionis Horn 75-130 (part)[49]
  s.dimorphicus Dari. 33-161 Alas.-Cal.-
                    [Colo.
  s.stygicus Chd. 71-46 [49]   Alas.-B.C.-
                   [Lab.-Newf.
    tenuis || Csy. 20-186 (not 1918)[49]
    PATROBUS (s.str.)
septentrionis Dej. 28-29 Eur. Sib. Alas.-
  fossor || O. Fab. 80-190 [49]   [Newf.
  hyperboreus Dej. 28-30 [49]
  fossifrons || Chd. 71-44 [49]
  labradorinus Csy. 18-395 [49]
  minuens Csy. 18-396 [49]
  tenuis || Csy. 18-396 (not Csy. 1920)[49]
  tritus Csy. 20-186 [49]
  GEOPATROBUS Darl. 38-157
foveocollis Esch. 23-105
  s.foveocollis (s.str.)[49]   Alaska
    fossifrons Dej. 28-31 [49]
    foveicollis Chd. 71-45 (part)[49]
    septentrionis Horn 75-130 (part)[49]
  s.tenuis Lec. 50-207[49] Alas.-Colo.-Lab.-
    angusticollis Mannh.53-146[49] [N.Y.
    foveicollis Chd. 71-45 (part)[49]
    tenuis Chd. 71-46
    obtusiusculus Chd. 71-43 [49]
    septentrionis Horn 75-130 (part)[49]
  laeviceps Csy. 18-397 [49]
  insularis Csy. 18-397 [49]

     *   *   *

Trechus Clairv. 06-22
  TRECHUS (s.str.)
obtusus Er. 37-122 [51]   Eur. Wash.
  quadristriatus || Duft. 12-185 [50]
  tristis || Steph. 27-170 [50]
  ? laevis Steph. 32-384 [50]
  castanopterus Heer 41-120 [50]
chalybæus Dej. 31-17
  v.brachyderus Jeann. 31-432 [52] Mt.Wy.
  v.californicus Mots. 45-347[52] Alas.-Ore.
  v.tahoensis Csy. 18-407 [52]   Cal. Nev.

[51] Hatch—33.
[52] Csiki—Col. Cat. p. 126.

coloradensis Schffr. 15-48
  v.arcticollis Jeann. 31-428 [52]   Ida.
  v.gravidulus Jeann. 31-429 [52]   N.M.
  ovipennis Mots. 45-348
  v.conformis Jeann. 27-188 [52]

Lasiotrechus Ganglb. 92-187
*   (*Blemus* || Redt. 58-67)[59]
discus Fabr. 01-207 [53] [54] Eur. Asia, Que.

PTEROSTICHINI

Stereocerus Kby. 37-34 [55]
Cylindrocharis Csy. 18-326
* grandiceps Lec. 48-336   Tenn.N.C. Ga.
  rostrata of Csy. 18-327 [58]
rostrata Newm. 38-387   Me.-Tenn.
  sulcatula Csy. 18-327 [56]
  piceata Csy. 18-327 [56]

Pterostichus
  *ORSONJOHNSONUS Htch 33-119
johnsoni Ulke 89-59 [51]   Ore. Wash.
  OMASEUS Steph. 28-67
vulgaris Linn. 58-415   Eur. B.C.-Ore.
  melanarius Ill. 98-163 [51] [57] [58]
  *HYPHERPES Chd. 38-8 [59]
tuberculo-femoratus Hatch 36-701 Ore.
castaneus Dej. 28-326   Alas.-Ore.
novellus Csy. 13-102   B.C.-Cal.
metlakatlæ Csy. 13-102   B.C.
terracensis Csy. 24-68   B.C.
stoicus Csy. 24-68   B.C.
amethystinus Mannh. 43-201 Alas.-Cal.
scutellaris Lec. 73-312   Cal.
ecarinatus Hatch 36-702   Ida. Wash.

     *   *   *

* LEPTOFERONIA Csy. 18-336 [59]
oregonus Csiki 30-582   Wash. Ore.
  longicollis Lec. 52-239 [59]
infernalis Hatch 36-705   Ore.
fenyesi Csiki 30-582   Cal.-Ore.
  ovicollis Schffr. 10-393 [59]
sphodrinus Lec. 63-10   Ida. Wash.
fuchsi Schffr. 10-392   Cal.
  fugiens Csiki 30-582
  fugax Csy. 18-337 [59]
  larvalis Csy. 18-337 [59]
  humilis Csy. 13-128 [59]
horni Lec. 73-313   Cal.
idahoensis Hatch 36-706   Ida.
angustus Dej. 28-328   Cal.
  crucialis Csy. 13-126 [59]
inanis Horn 91-32   Wash.-Cal.-Nev.
idahoæ Csiki 30-582   Ida.
  elongatus || Schffr. 10-365 [59]
beyeri VanD. 25-71   Mont.

[53] Chagnon—34.
[54] Chagnon—35.
[55] Transferred from Amarini by Leech—35.
[56] Nicolay & Weiss—34.
[57] Numerous other synonyms listed in Csiki—Col. Cat. p. 112.
[58] Leech—35.
[59] Revision of subgenus, Hatch—36.

termitiformis VanD. 25-74   Ore.
falli VanD. 25-73   Cal.
caligans Horn 91-33   Cal.
    ITHYTOLUS Bates 84-277
arizonicus Schffr. 10-393 [60]

Anilloferonia VanD. 26-115
* testacea VanD. 26-116   Wash.
lanei Hatch 35-116   Ore.

Euferonia Csy. 18-365 [61]
* iripennis Nic. & Weiss 34-200   S.C.
stygica Say 25-41   Ont.-N.C.-La.-Ia.
    bisigillatus Harris 28-123 [61]
    rugicollis Hald. 43-300 [61]
    picipes Newm. 38-377 [61]
    quadrifera Csy. 18-366 [61]
    proba Csy. 13-133 [61]
    ingens Csy. 18-367 [61]
    umbonata Csy. 18-368 [61]
    subaequalis Csy. 18-368 [61]
    v.*vapida* Csy. 13-134 [61]   N.H.-N.Y.
lachrymosa Newm. 38-386   Me.-N.C.-O.
washingtonensis Nic.&Weiss 34-203 N.H.
    v.*rufitarsis* Nic.&Weiss 34-204 [61] N.C.
coracina Newm. 38-386   Newf.-Va.-
    moerens Newm. 38-387 [61]   [L.Sup.
    adjuncta Lec. 52-245 [61]
    venator Csy. 20-189 [61]
    lacustris Csy. 24-71 [61]
    v.*roanica* Csy. 20-188 [61] Md.-N.C.-Ky.
    strigosula Csy. 24-72 [61]
    v.*erebea* Csy. 13-134 [61] Me.-Pa.-Wis.
    ludibunda Csy. 20-189 [61]
relicta Newm. 38-387 N.Y.-N.C.-L.Sup.
    protensa Lec. 63-12 [61]   (not La.)
flebilis Lec. 52-245   L.Sup.

Cryobius Chd. 38-11 [62]
mandibularis Kby. 37-31   Arctic N.A.
    arcticola Chd. 68-339 [63]
    hudsonicus Lec. 63-11 (Pterost.)[63]

Melanius Bon. 10-tab. syn. [61][62]
*   (*Omaseus* Leng, not Steph.)[61]
ebeninus Dej. 28-307   N.Y.-Fla.-Tex.
    acutangulus Chd. 43-771 [61]
caudicalis Say 23-56   N.J.-Va.-Ore.
    nigrita Kby. 37-32 [61]
    agrestis Bland. 65-381 [61]
    brevibasis Csy. 24-73 [61]
luctuosus Dej. 28-284   Newf.-Va.-Ill.
    abjectus Lec. 52-243 [61]
    hamatus Harr. 28-123 [61]
    confluens Csy. 24-73 [61]
    testaceus Csy. 24-74 [61]
    v.*tenuis* Csy. 24-73 [61]
corvinus Dej. 28-281   N.Y.-Ga.-L.Sup.
    subpunctatus Harr. 28-123 [61]
    tenebrosus Chd. 37-30 [61]
    aequalis Csy. 24-72 [61]

Dysidius Chd. 38-8 [61]
* purpuratus Lec. 52-242 [64]   N.J.-Va.-Ill.
    ?*parallelus* Mots. 59-147 [61]
    ohionis Csiki 30-638-[61]
    trinarius Csy. 18-377 [61]
mutus Say 23-44   N.E.U.S.
    carbonarius Dej. 28-283 [61]
    morosus Dej. 28-283 [61]
    picicornis Kby. 37-33 [61]
    pulvinatus Hausen 91-253 [61]
    stenops Hausen 91-252 [61]
v.egens Csy. 24-74 [61]   N.J. E.U.S.

Loxandrus
infimus Bates 82-87 [63]   Gua. Mex. Tex.

Amara
    CURTONOTUS Steph. 28-138
aulica Panz. 97-3 [66]   Eur. Asia, N.S.
    AMARA (s.str.)
familiaris Duft. 12-119 [67]   Eur. N.Eng.
    humilis Csy. 18-302 [67]
ænea DeG. 74-98 [67]   Eur. N.Eng.
    devincta Csy. 18-307 [68]

Platynus Bon. 09.[69]
    LEUCAGONUM Csy. 20-99 [69]
    AGONUM Bon. 09-tab. syn.[69][70]
pallipes Fabr. 01-187 [70]
    limbatus Say 23-49 [70][71]
    MELANAGONUM Csy. 20-111 [69]
    PARAGONUM Csy. 20-123 [69]
belleri Hatch 33-120 [69]   Wash.

• • •

    (*obscurus Hbst. 84-139*, not *N. Amer.*)[69]
    OXYPSELAPHUS Chd. 43-415 [69]
pusillus Lec. 54-39   Wash.-Can.-N.Y.
    americanus || Lec. 48-256 [69]
    oblongus || Ham. 88-61 [69]
    obscurus || Ham. 94-354 [69]

Colpodes
falli Dari. 36-152   Ariz.

Lachnophorus Dej. 31-28 [72]
    ARETAONUS Liebke 36-461 [72]
elegantulus Mannh. 43-215 Cal.-Tx.-C.R.
    mediosignatus Mén. 44-62 [72]
sculptifrons Bates 78-604 [72] Tx.-Mx.-Nic.

    COLLIURINI [74]

---

[64] This name was invalidated (according to some interpretations of the Rules) by being at one time transferred to Pterostichus where there is an older use of the name.
[65] Darlington—38.
[66] Fall—34.
[67] Darlington—36.
[68] Also numerous other synonyms in Csiki—Col. Cat. p. 104.
[69] Rearrangement of subgenera. Hatch—33.
[70] Buchanan—39 (cites Agonum as genus).
[71] Fall—33.
[72] Revision of genus, Liebke—36.
[73] Leng Catalog.
[74] Revision of tribe, Liebke—37.

[60] Placed in Leptoferonia by Hatch—36, but transferred here by Darlington—36.
[61] Revision of genus, Nicolay & Weiss—34.
[62] As subgenus of Pterostichus by Csiki—Col. Cal. p. 112.
[63] Brown—37.

Colliuris DeG. 74-79 [74]
* ODACANTHELLA Lbk. 30-658 [74]
pennsylvanica Linn. 58-620 U.S.
lengi Schffr. 10-395 Ariz.
CALOCOLLIURIS Lbk. 37-55 [74]
ludoviciana Sallé 49-297 Fla.-La.-Yuc.

Calybe Cast. 34-92 [74]
* EGA Cast. 34-93 [74]
sallei Chevr. 39-308 Fla.-La.-Nic.
lætula Lec. 49-173 Colo.-Mex.-Gua.

Leptotrachelus Latr. 29-371 [74]
* pallidulus Mots. 64-218 Tex.
dorsalis Fabr. 01-229 U.S. Mex. Cuba
depressus Blatch. 23-15 Fla.

Comstockia VanD. 18-179[74]
* subterranea VanD. 18-182 Tex.

* * *

Zuphium Latr. 06-198 [75]
* magnum Schffr. 10-396 Tex.
longicollis Lec. 79-62 Tex. Mex.
delectum Lbke. 33-469 Mass.
americanum Dej. 31-298 U.S.
mexicanum Chd. 62-314 Ariz. Mx.-C.R.

Pseudaptinus Cast. 35-50 [76]
* PSEUDAPTINUS (s.str.)[76]
lecontei Dej. 31-301 Fla. Ga. La.
tenuicornis Chd. 72-6 Tex. Mex.
tenuicollis Lec. 49-173 Cal. Tex.
THALPIUS Lec. 51-174 [76]
pygmæus Dej. 26-460 Fla. La. Cuba
högei Bates 82-77 Tex. Mex.
horni Chd. 72-3 Cal.
rufulus Lec. 49-173 Cal.

* * *

dorsalis Brullé 34-181 [77] D.C. Fla. La.
cubanus Chd. 62-252 [78] Fla. Cuba

Brachynus
BRACHYNUS (s.str.)[79]
affinis Lec. 48-204 [79] Ind.
alternans Dej. 25-316 [79] Ind.
alternans || Lec. 48-198 [79] U.S.
americanus Lec. 44-48 [79] Ind.Fla.
ballistarius Lec. 48-199 [79] N.Y.Ind.N.M.
carinulatus Mots. 59-139[79] Cal.Ariz.Mx.
glabripennis Lec. 58-28 [79]
cinctipennis Chev. 34-163 [79] Ar.N.M.Mx.
conformis Dej. 31-427 [79] U.S.
patruelis Lec. 44-50 [79]
cordicollis Dej. 26-466 [79] U.S.
velox Lec. 48-206 [79]
conformis || Lec. [79]
?cephalotes Dej. 25-317 [79]
costipennis Mots. 59-138 [79] Ariz.Cal.
lecontei Mots. 59-39 [79]
cyanipennis Say. 23-143 [79] Fla.Ind.

deyrollei Laf. 41-42 [79] Tex.Ind.
strenuus Lec. 44-48 [79]
a.tormentarius Lec. 48-200 [79] S.St.
[W.St.
distinguendus Chd. 68-287 [79] N.A.
fidelis Lec. 62-524 [79] Cal.-N.M.
fumans Fab. 81-307 [79] N.Y.-Fla.
librator Dej. 31-425 [79]
cyanopterus Lec. 44-49 [79]
perplexus || Lec. 48-203 [79]
sufflans Lec. 48-204 [79]
a.similis Lec. 48-199 [79] N.Y.
gracilis || Blatch. 10-160 [79] Ind.
janthinipennis Dej. 31-412 [79] U.S.
kansanus Lec. 62-524 [79] Kans.Ariz.
lateralis Dej. 31-424 [79] U.S. C.A. S.A.
leptocerus Chd. 68-296 [79] N.A.
leucoloma Chd. 68-301 [79] Cal.
limbiger Chd. 76-81 [79] Cal. Mex.-S.A.
medius Harris 28-117 [79] N.Eng. Ind.
minutus Harris 28-117 [79] N.Y. Ind.
pumilio Lec. 48-208 [79]
ovipennis Lec. 62-525 [79] U.S.
cephalotes || Lec. 48-205 [79]
phæocerus Chd. 68-300 [79] N.A.
puberulus Chd. 68-294 [79] Fla.
pulchellus Blatch. 10-161[79] Ind. Fla.
quadripennis Dej. 25-316 [79] S.St. W.St.
neglectus Lec. 44-49[79]
rejectus Lec. 62-525 [79] M.St. W.St.
cordicollis || Lec. 48-206 [79]
rugipennis Chd. 68-297 [79] U.S.
stenomus Chd. 68-291 [79] U.S.
stygicornis Say 34-415 [79] Kan. Fla.
oxygonus Chd. 43-714 [79]
sublævis Chd. 68-293 U.S.
tenuicollis Lec. 44-49 [79] U.S.
texanus Chd. 68-299 U.S.
tschernikhi Mannh. 43-184 [79]
puncticollis Lec. 58-28 [79]
viridipennis Dej. 31-426 [79] U.S.
viridis Lec. 44-49 [79]
lecontei Lec. 44-49 [79]
a.perplexus Dej. 31-426 [79] U.S.

Miscodera Esch. 30-63 [80]
* insignis Mannh. 52-296 Alas. Wash.
arctica Payk. 00-85 Eur. Sib.
s.erythropus Mots. 44-75 [80] Sib. Alas.-
americanus Mann. 53-134[80] [N.Y.
hardyi Chd. 61-525 [80]

Chlænius
crestonensis Brown 33-43 B.C.
leucoscelis Chev. 34-71
s.leucoscelis (s.str.)[81] Wash.-Colo.-
monachus Lec. 51-180 [81] [Guat.
chlorophanus || Lec. 48-435 [82]
s.cordicollis Kby. 37-22 [81]
[E.Can.-L.Sup. Va.

[75] Revision of genus, Liebke—33.
[76] Revision of genus, Liebke—34.
[77] Leng Catalog.
[78] Darlington—35.
[79] Synonymy revised by Csiki—33.
[80] Revision of genus, Hatch—33.
[81] Darlington—34.
[82] Csiki—31.

Selenophorus
discopunctatus Dej. 29-92 S.A. W.I. Fla.
chokoloskei Leng. 15-596 [82]
cuprinus Dej. 29-96 [84]
harpaloides Reiche 43-142 [84]
aeratus Reiche 43-142 [84]

Stenomorphus Dej. 31-696 [85]
• (Agaosoma Mén. 44-63) [85]
convexior Notm. 22-103   Ariz. Mex.
californicus Mén. 44-63   Kan.-Mex.-
rufipes Lec. 59-59 [85]   [Cal.-L.Cal.
batesi Csy. 14-168 [85]
scolopax Csy. 14-169 [85]
arcuatus Csy. 24-122 [85]
parallelus Csy. 24-122 [85]

Agonoderus
lecontei Chd. 68-14
pallipes of authors (not Fab.) [86] [87]

Pseudomorpha Kby. 25-98 [88]
• (Heteromorpha Kby. 25-109) [58]
(Axinophorus Dej. & Bois. 29-60) [88]
(Drepanus Dej. 31-434) [88]
alutacea Notm. 25-17   N.M.
angustata Horn 83-274   Ariz. N.M.
behrensi Horn 70-76   Cal. Mex.
castanea Csy. 09-278   Ut. Cal.
champlaini Notm. 25-20   Ariz. Cal.
consanguinea Notm. 25-18   Cal. Ariz.
cronkhitei Horn 67-151   Cal.
cylindrica Csy. 89-40   Tex.
excrucians Kby. 25-101   Ga. La. Ala.
lecontei Dej. & Bois. 29-176 [88]
falli Notm. 25-15   Cal.
hubbardi Notm. 25-15   Ariz.
ruficollis Csy. 24-148   La.
schwarzi Notm. 25-21   Ariz.
tenebroides Notm. 25-16   Ariz.
vandykei Notm. 25-18   Ariz.
vicina Notm. 25-17   Cal.
vindicata Notm. 25-19   Utah

HALIPLIDÆ

Haliplus Latr. 02-77 [1]
•   HALIPLUS (s.str.) [1]
strigatus Robts. 13-110   Man.-B.C.-Wyo.
robertsi Zimm. 24-73   Colo.-B.C.-Cal.
pallidus || Robts. 13-109 [1]
dorsomaculatus Zimm. 24-75 Cal. Colo.
distinctus Wallis 32-17   B.C.
hoppingi Wallis 32-19   B.C.
longulus Lec. 50-211   N.A.

[82] Durlington—35.
[84] Csiki—32.
[85] Revision of genus. Darlington—36.
[86] Buchanan—39.
[87] Fall—33.
[88] Csiki—33.
[1] Revision of genus. Wallis—32.

immaculicollis Harr. 28-164   N.A.
americanus Aube 38-26 [1]
impressus Kby. 37-65 [1]
ruficollis || Cr. 73-385 [1] .
blanchardi Robts. 13-108 Ct.-Minn.-La.
    PARALIAPHLUS Guignot 28-138 [1]
borealis Lec. 50-212 L.Sup. Minn. Man.
lewisii Cr. 73-384   Tex.
ohioensis Wallis 32-27   O. Ill.
minor Zimm. 24-192   Tex.
annulatus Robts. 13-107   Fla. S.C.
confluentus Robts. 13-106   Fla. S.C.
triopsis Say 25-106 N.Eng.Fla.N.M.Ont.
pantherinus Aube 38-29 E.N.A.Minn.La.
deceptus Math. 12-166   Tex. N.M.
suturalis Robts. 13-96 [1]
punctatus Aube 38-32   Fla.-Tex.
mutchleri Wallis 32-38   Fla.
leopardus Robts. 13-98   Mass.-S.C.
pseudofasciatus Wallis 32-41 S.C. Kan.
•   LIAPHLUS Guignot 28-138 [1]
fasciatus Aube 38-30 N.Eng.-S.C.-Kan.
connexus Math. 12-164 N.S.-Mass.-Minn.
mimeticus Math. 12-168 Pac.Cst. Mex.
rugosus Robts. 13-102   Cal.
apostolicus Wallis 32-46   Minn.
cribrarius Lec. 50-202 Lab.-Vt.-Minn.
canadensis Wallis 32-51   Mass.-B.C.
nitens Lec. 50-212   Mich.-Wis.-Tex.
subguttatus Rbts. 13-101 N.S.-N.Eng.-
                                  [B.C.
salinarius Wallis 32-56   B.C.
vancouverensis Math. 12-168   B.C.
salmo Wallis 32-61   Alb.
leechi Wallis 32-63   B.C.
columbiensis Wallis 32-66   B.C.
ungularis Wallis 32-68   B.C.
gracilis Robts. 13-102   Ore. Cal.
cylindrious Robts. 13-102   Cal.
tumidus Lec. 80-166   Tex. Cal.
concolo Lec. 52-201   Cal.

DYTISCIDÆ

Hygrotus Steph. 28-46 [2]
        COELAMBUS Thoms. 60-13 [2]
curvipes Leech 38-84   Cal.

Hydroporus Clairv. 06-182 [3]
        HYDROPORUS (s.str.) [3]
        HETEROSTERNUS Zimm. [3]
        DERONECTES Shp. [3] [4]
brodei Gellerm. 28-63 [5]
        STICTOTARSUS Zimm. [3]
        POTAMONECTES Zimm. [3]
        (Potamodytes Zimm.) [3]
        (Deronectes of Fall) [3]
mathiasi Hatch 33-22   Wash.

[2] Balfour-Browne—34.
[3] Rearrangement of subgenera. Hatch—33.
[4] Treated as valid by Balfour-Browne—34.
[5] Hatch—33.

OREODYTES Seidl.[6]
scitulus Lec. 55-295   B.C.-N.Y.-Newf.
bisulcatus Fall 23-115   Cal.
crassulus Fall 23-119   Mont. Ut. Wash.
obesus Lec. 66-365   Cal.-B.C.
angustior Hatch 28-221   Wash.
congruus Lec. 78-452   N.M.-Mont.-B.C.
abbreviatus Fall 23-117   Cal. Wash.
picturatus Horn 83-283   Nev. Cal.
subrotundus Fall 23-118   Cal. Wash.
snoqualmie Hatch 33-26   Wash.
hortense Hatch 33-27   B.C. Wash.
lævis Kby. 37-67   L.Sup. H.B.T.
    duodecimlineatus Lec. 50-214 [6]
semiclarus Fall 23-113   Cal. Y.T. Colo.
    yukonensis Fall 26-138 [6]
    recticollis of Hatch [6]
recticollis Fall 26-140   Alaska
alaskanus Fall 26-139   Alaska
rainieri Hatch 28-220   Wash.
    kincaidi Hatch 28-221 [6]
    GRAPTODYTES Seidl.[3]
    STICTONOTUS Zimm.[3]
    NEBRIOPORUS Reg.[3]

* * *

barbarensis Wallis 33-262   Cal.
planiusculus Fall 23-58   Que. B.C.
    brumalis Brown 30-235 [7]
    compertus Brown 32-4 [7]
    falsificus Brown 33-44 [7]
lapponum Gyll. 08-532 [8]
    [Alas.-Man. Eur. Sib.
melanocephalus Marsh 02-423[9] Eur. Sib.
    [Alas.-L.Sup.-Greenl.
    morio Gemm. & Har. 68-437 [9]
browni Wallis 33-261   Wash.
edwardsi Wallis 33-261   Wash.

Gozis **Agaporus** Zimm. 19-147
&t.   latens Fall 37-10   N.H. Mass. N.Y.
22)

Agabus
    GAURODYTES Thoms. 60-57 [10]
    falli Guignot 35-38
      sharpi || Fall 22-19 [10] [11]
    ERIGLENUS Thoms. 60-55 [12]
hudsonicus Leech 38-123   Man.

* * *

colymbus Leech 38-125   Man.
browni Leech 38-126   Man.
pseudoconfertus Wallis 26-90 [13]
    gelidus || Fall 26-142 [14] [13]
kenaiensis Fall 26-141   Alas. Man.
    palustris Wallis 26-92 [7]
vancouverensis Leech 37-146 B.C. Wash.
irregularis Mannh. 53-159 [14]   Alaska
hypomelas Mannh. 43-221 [14] Alas.-Wash.
audeni Wallis 33-270   B.C.
minnesotensis Wallis 33-268   Minn.

Ilybius
    denikei Wallis 33-271   Ont.

Copelatus Er. 32-38
    (LIOPTERUS Steph. 35-393)[15]

Rantus Bois. & Lac. 35-309 [16] [17] [18]
    (RHANTUS of authors)[16] [17] [18]
maculicollis Aube 38-245 [10] [19]   N.A.
hoppingi Wallis 33-272 B.C. Cal. Wash.
binotatus Harr. 28-164 [20] ·   N.A. C.A.
gutticollis Say 34-442 [20]   Ariz.
suturellus Harris 28-164[20] Eur. Sib. N.A.
    subopacus Menetr. 06-175 [20]
zimmermanni Wallis 33-274 Man. Que.
                [B.C.
    bistriatus of authors (not Bergst.)[20]

Hoperius Fall 27-177 [21] [11]
* planatus Fall 27-178 [21] [11]   Ark.

Dyticus Linn. 58-411 [22]
    (DYTISCUS of authors)[22]
habilis Say 34-441 [23]   Mex. Tex.

Hydaticus
    modestus Shp. 82-650   N.A.
    americanus Shp. 82-654 [17]
    stagnalis of authors (not Fab.)[17]
    (laevipennis Thoms., not N. Amer.)[17]
cinctipennis Aubé 38-191 [17]

Graphoderus Dej. 33-54 [24] [25] [26]
* (GRAPHODERES Thoms. 60-38)[24]
liberus Say 25-160 [25]   Mich.-Mass.-Fla.
    rugicollis Kby. 37-73 [25]
    brunnipennis Aube 38-203 [25]
    thoracicus Harris 28-156 [26]
fasciatocollis Harris 28-156[25] B.C.-Mass.
    cinereus Horn (not Linn.)[25]
perplexus Shp. 82-695 [25]   N.A.
    elatus Shp. 82-695 [25]
    cinereus of authors (not Linn.)[25]
    zonatus Zimm. (not Hoppe)[25]
manitobensis Wallis 33-276 [26]   Man.
occidentalis Horn 83-281 [25]   Cal.
    austriacus Zimm. (not Sturm.)[25]

---

[15] Mequignon—37, may be subgenus; proper date is 1835 in spite of statement to contrary by Strand.
[16] Balfour-Browne—35.
[17] Wallis in litt.
[18] Compiler is of opinion that the Rules amply justify the change of spelling to Rhantus; see remarks elsewhere.
[19] Hatch—28.
[20] Wallis—33.
[21] Leech in litt.
[22] Balfour-Browne—35, gives this as proper synonymy but does not accept it because the change was made four years late by another writer!
[23] Darlington—38.
[24] These names are reversed by Balfour-Browne —35, but Graphoderus was amply validated by Dejean in 1833.
[25] Revision of genus, Wallis in litt.
[26] Credited by Guignot—31 and Wallis in litt. to Aube 38-156.

---

[6] Revision of subgenus, Hatch—33.
[7] Fall—34.
[8] Possibly subsp. labradorensis Fall, Brown—37.
[9] Brown—37, ascribes these names to Gyll. & Aube respectively.
[10] Guignot—35.
[11] Fall in litt.
[12] Leech—38.
[13] Synonymy reversed by Fall—34.
[14] Leech—37.

GYRINIDÆ

Gyrinus
hoppingi Leech 38-59   B.C.

LIMNEBIIDÆ [27]

Ochtheblus [28]
insulanus Brown Correct loc. Vanc.Id.[27]
mimicus Brown 33-45   B.C.

HYDROPHILIDÆ

Helophorus
arcticus Brown 37-109   Baffln Id.
  HELOPHORUS (s.str.)
lecontei Knisch 24-88 [29]   Cal.
  obscurus || Lec. 52-210 [29]
ventralis Mots. 60-105 [29]   N.Y.
  obsoletesulcatus Mots. 60-106 [29]
  obscurus || Wickh. 95-183 [29]
granularis Linn. 61-214 [29]   Eur. N.A.
  fpusillus Mots. 60-106 [29]

Berosus
  ENOPLURUS Hope 38-128 [29]
undatus Fabr. 92-185 [29]   Tex. Mex. W.I.
  emarginatus Horn 73-120 [29]   [S.A.
  flavipes Shp. 87-766 [29]
  guadelupensis Fleut. & Salle 89-376[29]

Hydrophilus
occultus d'Orch. 33-310   N.Y. Tenn.

Enochrus
  LUMETUS Zaitz. 08-385 [29]
pygmæus Fabr. 92-186 [29]
  nebulosus Say 24-277 [29]   W.I.
P. 19, First supplement, under 19289,
  for 2484 read 2884. [27]

SILPHIDÆ

NECROPHORINI

Necrocharis Port. 23-68 [1][3][3]
* carolinensis Linn. 71-530 [4]   Pa.-Fla.-
  mediatus Fahr. 01-334 [2] [Ariz.-Neb.
  a.scapulatus Port. 23-142 [3]   N.C. Fla.
  a.dolosus Port. 23-142 [2]
  a.mysticallis Angell 12-307 [2]   Ariz.

[27] Leech in litt.
[28] Left in Hydrophilidæ by Brown—33.
[29] d'Orchymont—34.
[1] Semenov-Tian-Shanskij—32.
[2] Hatch—Col. Cat. p. 95.
[3] Placed as subgenus of Nicrophorus by Hatch, Col. Cat. p. 95.
[4] Hatch & Rueter—34.

Nicrophorus Fahr. 75-71 [2][4]
* (Dermestes || Geoffr. 62-93)[2]
  (Necrophorus Fahr. 01-333)[2][5]
  (Cyrtoscelis Hope 40-149)[2]
  Acanthopsilus Port. 14-223)[2]
  EUNECROPHORUS Sem.
               [32-152
americanus Cliv. 90-10 [1]   N.S.-Minn.-
  virginicus Fröl. 92-123 [2] [Tex.-Fla.
  grandis Fab. 92-247 [2]
germanicus Linn. 58-359 [2]   Europe
  a.bipunctatus Kr. 80-117 [2]   Eur. ?Cal.
  NECROPTER Sem. 32-154 [4]
orbicollis Say 25-77 [1] Me.-Fla.-Kan.-Alb.
  halli Kby. 37-98 [2]
  quadrisignatus Cast. 40-1 [2]
sayi Cast. 35-2 [1]   .   N.B.-Va.-Alb.
  lunulatus Gistl 48-49 [2]
  lunatus || Lec. 53-277 [2]
  luniger Har. 68-104 [2]
humator Cliv. 90-8- [2]   Eur. Asia
  grandior Angell 12-307 [2]   Cal.
defodiens Mannh. 46-513 [1][4]   Newf.-N.J.-
  hebes Kby. 37-96 [2]   [Cal.-Alas.
  pygmaeus Kby. 37-98 [2]
  vespilloides of Lec. 66-367 [2]
  a.humeralis Hatch 27-7 [2]
  a.binotatus Port. 26-236 [2]   Cal.
    plagiatus || Mots. 69-252 [2]
  a.lateralis Port. 03-330 [2]   Cal.-Wash.
  a.conversator Walk. 66-320 [2] Cal. Ore.
    pollinctor of Lec. 54-19 [2]
  a.kadjakensis Port. 26-236 [2]   Alaska
  a.mannerheimi Port. 24-293 [2]   Cal.
investigator Zett. 24-154 [1][4]   Eur.Asia,
  ruspator Er.37-225[2] [Alas.-Cal.-?Va.
  melsheimeri Kby. 37-97 [2]
  infodiens Mannh. 53-170 [2]
  confossor Mots. 60-126 [2]
  microcephalus Thoms. 62-9 [2]
  pustulatus of Horn 80-233 (part)[2]
  vestigator of Gyll. 27-308 [2]
  a.intermedius Reitt. 95-327[2] Eur. B.C.
               [Man.-Ore.-Ut.
  a.lutescens Port. 24-150 [2] N.M. Ariz.
  v.nigritus Mannh. 43-251 [2]   Cal. Ore.
  v.variolosus Port. 24-149 [2]
  s.maritimus Guer. 44-60 [2] Japan, B.C.
  aleuticus Gistel 48-190 [2]
  pollinctor Mannh. 53-169 [2]
  sibiricus Mots. 60-126 [2]
  a.particeps Fisch. 44-139 [2]
             [Alas. B.C. Asia
mexicanus Matth. 88-91 [1]   Mex. Cal.
hybridus Hatch & Ang. 25-216 [1][4] Wash.-
                   [Man.
  v.minnesotianus Hatch 27-5 [2]   Minn.
guttulus Mots. 45-53 [1][4]   Alas. Cal.-Colo.
  a.quadriguttatus Angell 20-90 [2]   Cal.
  a.vandykei Angell 20-90 [2]   Cal.
  a.ruficornis Mots. 69-352 [2]   Cal.
  a.hecate Bland || 65-382 [1][2] Alas.-N.M.
  a.disjunctus Port. 24-85 [2]
  a.rubripennis Port. 24-85 [2]   Cal.-Kan.
obscurus Kby. 37-97 [1][4] H.B.T.-Ut.-Ore.
  melsheimeri of Lec. 53-275 [2]

* This spelling accepted by Semenov-Tian Shanskij—32.

marginatus Fabr. 01-334 [1][4] Me.-Miss.-
  requiescator Gistel 48-190[2] [Ore.-C.A.
  a.cordiger Port. 24-84[2] N.Y. Ark.
tomentosus Web. 01-47[1] Me.-Ga.-Cal.-
  velutinus Fab. 01-334[2] [Man.
  a.angustefasciatus Port. 25-170[2]
  a.aurigaster Port. 25-170[2]
  STICTONECROPTER Sem. 32-154
pustulatus Hersch 07-271[1] Alas.-Newf.-
  bicolor Newm. 38-385[2] [Fla.-Cal.
  tarsus Mannh. 53-385[2]
  a.fasciatus Port. 24-86[2] Ky.-Que.-Fla.
  a.unicolor Port. 24-86[2] La.
vespillo Linn.58-359 Eur. Asia, Neb. Pa.
pulsator Gistel 48-190[2] "N.A."
vespilloides Hbst. 84-32[6] Eur. Japan
  s.defodiens Mannh. 46-513[6] Pac. Cst.
  mortuarum Fab. 92-248[6] [Japan
  a.hebes Kby. 37-96[6] Alas. Y.T.
  pygmaeus Kby. 37-98[6]
  humeralis Hatch 27-7[6]
  a.conversator Walk. 66-320[6] B.C.
  lateralis Port. 03-330[6]
  gaigei Hatch 27-356[6]

SILPHINI

Silpha
  THANATOPHILUS Leach 15-89[4]
trituberculata Kby. 37-101[4]
lapponica Hbst. 93-269[4]
  HETEROSILPHA Port. 26-83[4]
ramosa Say 23-193[4]
  XYLODREPA Thoms. 59-56
quadripunctata Linn. 58-359[a] Eur. Mass.
  OXELYTRUM Gistel 48-150
discicollis Brullé 40-75[b] Cal. N.M. Ariz.
  analis Chev. 43-26[a][b] [C.A. S.A.
  æquinoctialis Gistel 48-190[b]
  brasiliensis Dej. 33-118[b]
  cayennensis || Berg 01-328[b]

AGYRTINI

Agyrtes Fröl. 99-18[7]
* longulus Lec. 59-282[8] Cal.-Alas.
  similis Fall 37-29 Cal.

Pelatines Ckll. 06-240[7]
* (*Pelates* || Horn 80-244)[7]
  latus Mannh. 52-331[8] Alas.-Wash.-Alb.

Necrophilus Latr. 29-500[7]
* pettiti Horn 80-243[8] Can.-Ky.-Ind.
  subterraneus of Horn 68-125[8]
  hydrophiloides Mannh. 43-253[8] Alas.-
  ater Mots. 45-263[8] [Cal.

a Angell in litt.
b Junk Col. Cat. pars 95.
[6] Leech—37.
[7] Rearrangement of genera. Hatch & Rueter—34.
[8] Leng Catalog and supplements.

LYROSOMINI [4][9]

Pteroloma Gyll. 27-418[10]
* (*Adolus* Fisch. 28-242)[2]
  (*Holoonemis* Schill. 29-93)[2]
  nebrioides Brown 33-213 Alas. Alb.
  forsströmii of authors (not Gyll.)[11]

Apteroloma Hatch 27-12[4][12][10]
* (*Pteroloma* of authors)[4]
  caraboides Fall 07-235[8] Wash.-Cal.
  tenuicornis Lec. 59-84[8] Wash.-Cal.
  tahoeca Fall 27-136[8] Cal.

CATOPOCERINI [4]

(*Pinodytini*)

Catopocerus Mots. 69-351[4]
  (*Pinodytes* Horn 80-248)[4]
  (*Homaeosoma* Aust. 80-16)[2]
  ulkei Brown 33-215
  cryptophagoides of Horn 80-249[11]

LEPTODIRIDÆ [1]

(*Catopidae, Silphidae*)

Leptodirinæ [1]

Platycholeus Horn 80-251[1]
* leptinoides Cr. 74-77[1] Cal. Ore. Nev.
  opacellus Fall 09-133[1] Cal.

Catopinæ [1]

CATOPINI

Nemadus Thoms. 67-351[1][2]
* pusio Lec. 59-282[1][2] Cal.-B.C.
  horni Hatch 33-194[2] Mass.-Ala.-Ia.
  pusio of Horn 80-262[1]
  parasitus Lec. 53-282[1][2] Que.-Va.-Tex.
  [Man.
  integer Fall 37-338 Mass.
  gracilicornis Fall 37-339 Mass.N.J.Man.
  obliquus Fall 37-339 Mass.-Pa.

Dissochætus Reitt. 84-39[1]
* brachyderus Lec.63-25[1] N.S.-N.Y.-Minn.
  arizonensis Hatch 33-197 Ariz.
  oblitus Lec. 53-282[1] Md.-Fla.-Ill.
  decipiens Horn 80-257[1] Wash.

Echinocoleus Horn 85-136[1]
* setiger Horn 85-136[1]

[9] Van Dyke—28.
[10] Definitely not Staphylinidæ, Brown—33, Szekessy—36, Van Dyke—28, the compiler, etc.
[11] Brown—33.
[12] Considered a synonym of Pteroloma, Brown—33.
[1] Hatch—38.
[2] Revision of genus, Fall—37.

Ptomaphagus Ill. 98-88 [1]
* californicus Lec. 53-281 [1]   Cal.
nevadicus Horn 80-263 [1] Nev. Colo. Kan.
schwarzi Hatch 33-203 [1]   Fla.
texanus Melander 02-329 [1]   Tex. Colo.
fisus Horn 85-137 [1]   Ariz. Cal.
consobrinus of Horn 80-262 (part) [1]
cavernicola Schw. 98-58 [1]   Mo.
latior Hatch 33-204   Cal.
ulkei Horn 85-137 [1]   Md. Va. N.C.
piperi Hatch 33-205   Wash. Cal.
consobrinus Lec. 53-281 [1]   Mass.-Fla.-
strigosus Lec. 53-281 [1]   [Ckla.-Mich.
lecontei Murray 56-459 [1]

Adelops Tellk. 44-318 [1]
* mitchellensis Hatch 33-208   N.C.
lödingi Hatch 33-209   Ala.
valentinei Jeann. 33-252 [3]   Ala.
batchi Jeann. 33-252 [3]   Tenn.
hirtus Tellk. 44-318 [1]   Ky.

Catops Payk. 98-342 [1]
*   SCIODREPA Thoms. 62-66 [1]
gratiosus Blanch. 15-294 [1] Me.-Va.-Wash.
alsiosa of Blatch. 10-294 [1]
alsiosus Horn 85-136 [1]   Alas.-N.S.-N.Y.
simplex Say 25-184 [1] Alas.-Lab.-Cal.-Ga.
luridipennis Mannh. 53-176 [1]
americanus Hatch 28-201 [1]   N.B.-N.C.-
   [N.M.-Minn.
clavicornis || Lec. 53-281 [1]
basilaris Say 23-194 [1] Me.-N.C.-Cal.-Alb.
spenciana Kby. 37-108 [1]
cadaverinus Mannh. 43-82 [1]
brunnipennis Mannh. 53-176 [1]
egenus Horn 80-256 [1]   Alas.-Cal.
hornianus Blanch. 15-294 [1] Que.N.C.B.C.
SCIODREPOIDES Hatch 33-224 [1]
(*Sciodrepa* of Ganglb.) [1]
terminans Lec. 50-218 [1]   B.C.-N.B.-Ill.
fumata of Hatch 27-15 [1]

Prionochæta Horn 80-260 [1]
* opaca Say 25-184 [1]   Que.-N.C.-Ark.-
   [Minn.

Catoptrichus Murray 56-461 [1]
* frankenhæuseri Mannh. 52-332 [1]
   [Alas.-Wash.

Coloninæ [1]

Colon Hbst. 97-224 [1]
*   CCLCN (s.str.) [1]
paradoxum Horn 80-270 [1]   Pa. D.C.
CURVIMANON Fleisch. 09-246 [1]
bidentatum Sahlb. 17-95 [1] Eur. B.C. Alb.
   [N.J.
MYLOECHUS Latr. 07-30 [1]
hubbardi Horn 80-270 [1]   Mich. Tenn. O.
productum Hatch 33-229   Mich.
excisum Hatch 33-229   N.C.
kincaidi Hatch 33-229   Wash.

[1] Hatch—33.
[3] Placed in genus Ptomaphagus, Jeannel—33.

dentatum Lec. 53-282 [1]   N. J. Que. O.
putum Horn 80-272 [1]   N.Y.
putatum Hatch 28-225 [1]
rufum Hatch 33-230   W.Va.
discretum Hatch 33-230   B.C.
rectum Hatch 33-231   W.Va.
pribilof Hatch 33-231   Alaska
schwarzi Hatch 33-232   N.C. Mich.
magnicolle Mannh. 53-177 [1]
   N.Y. O. Wis. Colo.
pusillum Horn 80-273 [1] Y.T. Que.-D.C.
thoracicum Horn 80-274 [1]   Va.
asperatum Horn 80-274 [1]   Cal.-Me.

LEIODIDÆ [4]

(*Liodesidae, Agathidiidae, Anisotomidae, Silphidae*)

LEIODINI [4]

Hydnobius Schm. 41-193 [4]
* matthewsii Cr. 74-74 [4]   Cal.-B.C. Mo.
pallidum || Say 34-91 [4]   [N.Y.
strigilatus Horn 80-280 [4]   B.C. Nev.
arizonensis Horn 85-138 [4]   Ariz.
obtusus Lec. 79-511 [4][5]   Colo. B.C.
luggeri Hatch 27-18 [4][5]   Alaska
longulus Lec. 79-511 [4][5] B.C.-Cal. Colo.
longidens Lec. 79-511 [4]
substriatus Lec. 63-25 [4][5]   N.S.-N.Y.-
curvidens Lec. 79-511 [4]   [Wash.
laticeps Notm. 20-27 [4][5]   N.Y.
pumilus Lec. 79-511 [4][5] Que. Colo. Cal.
latidens Lec. 79-512 [4][5]
kiseri Hatch 36-35   Wash.
lobatus Hatch 36-35   Wash.
femoratus Hatch 36-36   Wash. B.C.

    * * *

validus Brown 32-202 (Leiodes) [6]   Que.

Typhloleiodes Hatch 35-116
* subterraneus Hatch 35-116   Ore.

Triarthron Märk. 40-141 [4]
* lecontei Horn 68-131 [4]   Ore. Cal.
cedonulli Schauf. 82-43 [4]
pennsylvanicum Horn 83-284 [4]   Pa.

Leiodes Latr. 02-163 [4]
*   (*Anisotoma* Schmidt 41-143, not
   [Panz.) [4][5]
   (*Liodes* Reitt. 84-93) [4]
   LEIODES (s.str.) [4]
alternata Melsh. 46-103 [4]   Ga.
americana Knoch MS [4]
canadensis Brown 28-141 [4] Sask. Man.
oklahomensis Brown 28-142 [4]   B.C.
horni Hatch 29-36 [4]   Cal. Ore.
humeralis || Horn 80-285 [4]
valida Horn 80-285 [4]   N.H.-Colo.-B.C.
serripes Hatch 36-37   Wash.
merkeliana Horn 95-234 [4]   Wash.

[4] Hatch—29.
[5] Hatch—36.
[6] Hatch in litt.

assimilis Lec. 50-221 [4]   N.H.-Colo.-B.C.
punctatostriata Kby. 37-110 [4]   Alas.-B.C.
   indistincta Lec. 50-221 [4]   [N.H.
   laeta Mannh. 53-201 [4]
difficilis Horn 80-285 [4]   Cal.
collaris Lec. 50-221 [4]   N.H.-Cal.-B.C.
similis Fall 10-5 [4]   Cal.
curvata Mannh. 53-202 [4]   Alas.-Cal.
   morula Lec. 59-282 [4]
opacipennis Fall 10-5 [4]   Tex.
antennata Fall 10-6 [4]   Cal.
lateritia Mannh. 52-345 [4]   Alaska

PSEUDOHYDNOBIUS Ganglb.
                  [99-208 [5]
    (*Parahydnobius* Fleisch.
                 [08-18) [5]
conferta Lec. 66-367 [4] [5]   Ill. Pa.
paludicola Cr. 74-74 [4] [5]   Cal.
strigata Lec. 50-221 [4] [5] Mich. Colo. B.C.

OREOSPHAERULA Ganglb.
                 [96-181 [4]
obsoleta Melsh. 46-107 [4]   Atl.St.-Cal.
sculpturata Fall 10-6 [4]   Ariz.
puritana Fall 25-310 [4]   Mass.
fusciclava Fall 25-311 [4]   Cal.

Anogdus Lec. 66-369 [4] [7] [8]
* luggeri Hatch 27-17 [4]   Minn.
dissimilis Blatch. 16-93 [4] [7]   Fla.
capitatus Lec. 66-319 [4] [7]   Fla.

Cainosternum Notm. 21-148 [4]
* imbricatum Notm. 21-148 [4]   N.Y.

Colenis Er. 41-221 [4]
* (*Colensia* Fvl. 02-287) [4]
   (*Pseudoliodes* Port. 26-77) [4]
impunctata Lec. 53-284 [4]   Fla.-La.-Mich.

AGATHIDIINI [4]

Anisotoma Panz. 97-8 [4] [9]
* (*Pentatoma* || Schneid. 92-2) [4]
   (*Leiodes* || Schmidt 41-132) [4]
   (*Liodes* Er. 45-87) [4]
blanchardi Horn 80-298 [4] [7] Que.Mass.O.
geminata Horn 80-299 [4] [7] Que.-N.C.-Ill.
confusa Horn 80-299 [4] [7]   Nev.
insterstrialis Hatch 36-38 [10] [7] [11]   Wash.
globososa Hatch 29-56 [4] [7] Que.-Ky.-Nev.-
   globosa || Lec. 50-222 [4] [7]   [B.C.
inops Brown 37-198   N.B.-Ont.-N.H.
nevadensis Brown 37-199   Nev.
   bicolor || Horn 80-297 [7] [12]
expolita Brown 37-199   Ont.-Ga.-Mich.
   polita || Lec. 53-285 [7]
basalis Lec. 53-285 [4] [7]   Mich. Ill. Ind.
   dichroa Lec. 53-285 [7] [12]
amica Brown 37-201   Que. Ont. B.C.
obsoleta Horn 80-298 [4] [7]   N.B.-Va.
errans Brown 37-202   Que. Ont. B.C.
discolor Mels. 46-103 [4] [7] Can.-Va.-Mich.
   piceum Mels. 44-103 [7]

Stetholiodes Fall 10-4 [4]
* laticollis Fall 10-4 [4]   Ind.

Neocyrtusa Brown 37-161
* obsoleta Melsh. 46-107 [7]   D.C. N.C.
   blandissima Zimm. 69-250 [4] [13]
secreta Brown 37-163   Ont.
superans Fall 10-7 [7] [13]   Ont.-Alb.
puritana Fall 25-310 [7]   Ont. Mass.
potens Brown 32-205 [7]   Que. Ont.
insolita Brown 37-170   B.C. Sask.
?fusciflava Fall 25-311 [7]   Cal.
?sculpturata Fall 10-6 [7]   Ariz.

Lionothus Brown 37-170
* ulkei Brown 37-171   D.C.
   blandissima of Horn 80-294 [7]

Cyrtusa Er. 42-221 [7]
* luggeri Hatch 27-17 [7]   Que.-Mich.-B.C.

Cænocyrta Brown 37-172
* picipennis Lec. 63-25 [7] [13] Que.-D.C.-B.C.

Apheloplastus Brown 37-173
* egenus Lec. 53-284 [7] [13]   Que.-Ga.-Mich.
   impubis Zimm. 69-251 [7]

Isoplastus Horn 80-277 [4] [7]
* fossor Horn 80-295 [4] [7]   Que.-D.C. Mich.

Agathidium Panz. 97-13 [14]
*   AGATHIDIUM (s.str.) [14]
oniscoides Beauv. 05-160 [14] [15] Can.-Ga.-
   piceum Mels. 44-103 [4]   [Miss. Mex.
   globatile Lec. 78-598 [4]
rubellum Fall 34-105 [14]   N.C.
compressidens Fall 34-106 [14] N.C. N.H.
exiguum Melsh. 44-103 [14] [15]   Can.-Fla.-
   ruficorne Lec. 50-222 [4]   [Colo. Guat.
   minutum Melsh. MS [4]
dentigerum Horn 80-301 [14] [15]   Va. N.C.
alutaceum Fall 34-107 [14]   Alaska
californicum Horn 80-301 [14] [15]   Cal.-
                 [B.C. Guat.
depressum Fall 34-108 [14]   Alas.-Cal.-
                 [Ill.-Que.
jasperanum Fall 34-109 [14]   Alb. Y.T.
dubitans Fall 34-110 [14]   N.M.
revolvens Lec. 50-222 [14] [15]   Alas.-Cal.-
                 [Mich.
cavisternum Fall 34-111 [14]   B.C.
virile. Fall 01-219 [14] [15]   S.Cal.
conjunctum Brown 33-46 [14] [15] Nev.-B.C.
omissum Fall 34-113 [14]   Mont.
    NEOCEBLE Gozis 86-16 [14]
sexstriatum Horn 80-302 [14] [15] Cal. Nev.
bistriatum Horn 80-303 [14] [15] Cal. Nev.
estriatum Horn 80-303 [14] [15]   Colo.
parvulum Lec. 78-598 [14] [15]   L.Sup.
parile Fall 34-117 [14]   Cal.

---

[7] Brown—37.
[8] Placed near Cyrtusa by Brown—37.
[9] Credited to Illiger, Hatch—36.
[10] Hatch—38.
[11] As synonym of confusa. Brown—37.
[12] As variety by Hatch—29.

[13] Placed in Cyrtusa (*Zeadolopus* Broun =) by Hatch—29.
[14] Hatch—36.
[15] Fall—34.

rusticum Fall. 34-117 [14]   N.H. Mass.
lætum Fall 34-118 [14]   Cal.
contiguum Fall 34-119 [14]   Wash.
athabascanum Fall 34-119 [14]   Alb.
alticola Fall 34-120 [14]   N.H.
columbianum Fall 34-121 [14]   B.C.
rotundulum Mannh. 52-370 [14] [15] Alas-Cal.
brevisternum Fall 34-122 [14]   Cal.
atronitens Fall 34-122 [14]   Ill. D.C. Mo.
repentinum Horn 80-303 [14]   N.H. Mont.
politum Lec. 66-370 [14] [15]   Can.-R.I.-Mo.
maculosum Brown 28-145[14] [15] B.C. Wash.
    maulosum Brown 28-230 [4]
v.franciscanum Fall 34-126 [14]   Cal.
kincaidi Hatch 36-39   Wash.
pulchrum Lec. 53-286 [14] [15]   N.H. Ky.
                              [Wash. Cal.
    mandibulatum Mannh. 53-203 [4]
varipunctatum Hatch 36-40   Wash.
picipes Fall 34-130 [14]   Cal.
difforme Lec. 50-222 [14] [15]   N.H. Mich.
    canadensis Brown 30-89 [14] [15]

    CYPHOCEBLE Thoms. 59-59 [14]
angulare Mannh. 52-369 [14] [15]   Alas.-Cal.-
                              [Ariz. Colo.
concinnum Mannh. 52-370 [14] [15]   Alas.-
                              [Cal.-Ariz.-Colo.
    effluens Mannh. 53-202 [4]
municeps Fall 34-171 [14]   N.H.
    temporale || Fall 34-127 [14] [15]
assimile Fall 34-128 [14]   Ind. N.H.
mollinum Fall 34-128 [15]   N.H. Me.

Aglyptinus Ckll. 06-240 [4]
• (*Aglyptus* Lec. 66-369)[4]
(*Aglyptonotus* Champ. 13-65)[4]
lævis Lec. 53-284 [4]   E.Can.-Ga.-La.

STAPHYLINIDÆ

Trigonurus Muls. & Rey 47-515 [1]
• cælatus Lec. 74-48 [1]   Cal.
    rugosus Shp. 75-204 [1]
crotchi Lec. 74-48 [1]   Cal.-Alas.
    lecontus Shp. 75-205 [1]
    lecontei of authors [1]
edwardsi Shp. 75-205 [1]   Cal. Wash.
dilaticollis VanD. 34-179 [1]   Cal B.C.[2]
(*subcostatum Mäkl.*, belongs *in*
                              [*Lathrimaeum*)[1]

Thoracophorus
brevicristatus Horn 71-332 [1]   Mex. W.I.
                              [Fla. Ariz.

Olophrum Er. 37-622 [3]
• (*Lathrium* Lec. 50-21)[3]
obtectum Er. 40-865 [3]   Me.-Pa.
    emarginatum Er. 40-868 [3]
    rotundicolle || Say 34-464 [3]
latum Mäkl. 53-194 [3] [4]   Alaska
boreale Payk. 92-146 [5] [6] Eur. Bear Id.
                              [Que. N.W.T.
    marginatum Kby. 37-89 [3] [4]
parvulum Mäkl. 53-195 [3]   Alaska
marginatum Mäkl. 53-196 [3]   Alaska
consimile Gyll. 10-199 [3]   Eur. Sib.
                              [Alas.-N.Y.
    v.minor J. Sahlb. 76-424 [3]   ?N.A.
rotundicolle C.R. Sahlb. 27-281 [3]   Eur.
                              [Sib.Newf.-Mich.-N.W.T.
    convexicolle Lec. 50-21 [3]
convexum Mäkl. 53-195 [3]   Alaska

Orobanus Lec. 78-453 [7]
• montanus Mank. 34-121   Mont.
    mormonus Mank 34-122   Utah
    falli Mank 34-122   Cal.
densus Csy. 86-246 [7]   Cal.
rufipes Csy. 86-245 [7]   Cal.
simulator Lec. 78-453 [7]   Colo. B.C.

Boreaphilus Sahlb. 34-433
• americanus Notm. 18-188 [8]   N.J.
nearcticus Blair 33-95 [6]   No.Que.

Deleaster
dichrous Grav. 02-188 [7] [10]   Eur. Que.

Trogophlœus
fulvipes Er. 40-804 [1]   S.A. C.A. W.I.
    rubripennis Fvl. 63-440 [1] [Fla.-Cal.
    senilis Shp. 80-51 [1]   [Hawaii
    texanus Csy. 89-334 [1]

Oxytelus
insignitus Grav. 06-188 [1]   S.A. C.A.
    [W.I. N.A. Eur. Atl.Is. Pac.Is.
    americanus Mannh. 30-48 [1]
    pumilio Boh. 58-34 [1]

Bledius
philadelphicus Fall 19-26 [11]
    dissimilis || Fall 10-107 [11]
    falli Wend. 28-298 [11]

Paralispinus
exiguus Er. 40-830 [1]   S.A. C.A. W.I.
    laevigatus Kr. 59-188 [1] [Fla.-La.
    rufescens Lec. 63-59 [1]   [Orient
    rufus Fvl. 65-60 [1]   [Hawaii
    fauveli Shp. 76-392 [1]
    aruensis Fvl. 78-200 [1]
    pallescens Blackb. 85-126 [1]

[3] Revision of genus, Scheerpeltz—29.
[4] Voris in litt.
[5] Blair—33.
[6] Brown—37.
[7] Rearrangement of genus, Mank—34.
[8] Leng Catalog.
[9] Chagnon—34.
[10] Jules—35.
[11] Fall in litt.

[4] Hatch—29.
[14] Hatch—36.
[15] Fall—34.
[1] Blackwelder in MS.
[3] R. Hopping—36.

Stenus
  falli Scheerp. 33-1154   Alas. Y.T.
    frigidus ‖ Fall 26-59 [12]

PAEDERINI [13]

Thinocharis Kr. 59-142 [13]
    SCIOCHARIS Lynch 84-260 [13]
  carolinensis Csy. (Sciocharis)
  congruens Csy. (Sciocharis)
  nubipennis Csy. (Sciocharis)
    SCIOCHARELLA Csy. 05-151 [13]
  exilis Er. 40-627 [1]   S.A. C.A. W.I.
    atratula Lynch 84-265 [1]   [Fla. Ala.
    fragilis Shp. 86-574 [1]
    minuta Shp. 86-574 (Sciocharis) [1]
    delicatula Csy. 05-159 (Sciocharis) [1]
    pertenuis Csy. 10-188 [1]

Lithocharis Bois. & Lac. 35-431 [13]
    (*Arthocharis* Cam. 21-372) [13]
    (*Metaxyodonta* Csy. 86-29) [13]
    (*Sunius* Steph. 32-274, not Er.) [13]
    LITHOCHARIS (s.str.) [13]
  ochracea Grav. 02-59 [1]   S.A. C.A. W.I.
    [U.S. Eur. Afr. India, China, Pac.Is.
    rubricollis Grav. 06-138 [1]
    testacea Bois. & Lac. 35-432 [1]
    brunniceps Fairm. 49-290 [1]
    fastidiosa Fairm. & Germ. 61-438 [1]
    alutacea Csy. 86-30 [1]
    quadricollis Csy. 86-31 [1]
  simplex Csy.
  sonoricus Csy.
    PSEUDOMEDON Muls. & Rey
                       [78-122 [13]
  alabamæ Csy. (Pseudomedon)
  capitulus Csy. (Pseudomedon)
  clarescens Csy. (Pseudomedon)
  ruficollis Csy. (Pseudomedon)
  thoracicus Csy. (Pseudomedon)

Aderocharis Shp. 86-552 [13]
    ADEROCHARIS (s.str.) [13]
  corticinus Grav.

Stilomedon Shp. 86-565 [13]
    POLYMEDON Csy. 05-151 [13]
  tabacinum Csy. (Polymedon)

Neomedon Shp. 86-557 [13]
  arizonense Csy.

Hypomedon Muls. & Rey 78-122 [13]
    (*Chloëcharis* Lynch 84-259) [13]
    (*Euastenus* Fiori 15-10) [13]
    (*Hemimedon* Csy. 05-152) [13]
    Lena Csy. 05-189) [3]
    HYPOMEDON (s.str.) [13]
  angustum Csy. (Hemimedon)
  curtipennis Scheerp. 33-1259 [12]
    brevipenne ‖ Csy. 05-186 (Calo-
                          [derma) [12]

[12] Scheerpeltz—33.
[13] A generic revision of the Pæderini of the world, Blackwelder — 39. Species examined are listed with indication of their former position if different from the present, but the list is not complete. See original catalog for citations, localities, and synonymy of the species.

conjux Csy. (Caloderma)
continens Csy. (Caloderma)
contractum Csy (Caloderma)
debilicorne Woll. 57-194 [1]   S.A. C.A.
    [W.I. Fla.-S.C.-Tex. Eur. Asia.
  brevicornis Allard 57-747 [1]
  ægyptiaca Mots. 58-644 [1]
  occulta Waterh. 76-108 [1]
  rufula Lynch 84-259 [1]
  pallidus Fiori 15-10 [1]
discolor Csy. (Caloderma)
exile Csy. (Caloderma)
luculentum Csy. (Caloderma)
mobile Csy. (Caloderma)
molle Csy. '(Caloderma)
peregrinum Csy. (Caloderma)
pollens Csy. (Caloderma)
quadripenne Csy. (Caloderma)
reductum Csy. (Caloderma)
rufipes Csy. (Hemimedon)
tantillum Csy. (Caloderma)
testaceum Csy. (Lena)
    OLIGOPTERUS Csy. 86-12 [13]
    (*Micromedon* Csy. 05-153) [13]
    (*Medonella* Csy. 05-154) [13]
cuneicolle Csy. (Oligopterus)
filum Csy. (Oligopterus)
flexile Csy. (Oligopterus)
minutum Csy. (Sciocharis)
remotum Csy. (Oligopterus)
    CALODERMA Csy. 86-5 [13]
angulatum Csy. (Caloderma)
rugosum Csy. (Caloderma)
semibrunneum Csy. (Caloderma)
    TRACHYSECTUS Csy. 86-32 [13]
confluentum Say (Trachysectus)

Medon Steph. 32-273 [13]
    (*Oxymedon* Csy. 05-177) [13]
    MEDON (s.str.) [13]
americanum Csy.
rubrum Csy. (Oxymedon)
texanum Csy.
    TETRAMEDON Csy. 05-178 [13]
rufipenne Csy. (Tetramedon)
    PLATYMEDON Csy. 89-184 [13]
    (*Paramedon* Csy. 05-166) [13]
arizonicum Csy. (Paramedon)
boreale Csy. (Paramedon)
conforme Csy. (Paramedon)
consanguineum Csy. (Paramedon)
contiguum Csy. (Paramedon)
convergens Csy.
caseyi Scheerp. 33-1256 [12]
    debilis ‖ Csy. 05-174 (Paramedon) [13]
difforme Csy. (Paramedon)
distans Csy. (Paramedon)
explicans Csy.
gregale Csy. (Paramedon)
gulare Csy. (Paramedon)
helenæ Csy.
humboldti Csy. (Paramedon)
inquilinum Csy.
insulare Csy.
kernianum Csy. (Paramedon)
lacustre Csy.

languidum Csy. (Paramedon)
laticolle Csy. (Platymedon)
latiusculum Csy. (Paramedon)
lepidum Csy.
luctuosum Csy. (Paramedon)
malacum Csy. (Paramedon)
mimulum Csy. (Paramedon)
montanum Csy. (Paramedon)
nevadicum Csy. (Platymedon)
nitidulum Csy.
oriens Csy. (Paramedon)
pallescens Csy. (Paramedon)
pallidipenne Csy. (Paramedon)
puberulum Csy.
retrusum Csy. (Paramedon)
sinuatocolle Csy.
shastanicum Csy. (Paramedon)
sublestum Csy. (Paramedon)
subsimile Csy. (Paramedon)
tahoense Csy. (Paramedon)
vancouveri Csy. (Paramedon)
    MEDONODONTA Csy. 05-176 [11]
alutaceum Csy. (Medonodonta)

Orus Csy. 84-136 [13]
    ORUS (s.str.) [13]
boreellus Csy.
deceptor Csy.
distinctus Csy.
filius Csy.
fraternus Fall
caseyianus Scheerp. 33-1265 [12]   Cal.
   longicollis || Csy. 05-197 [12]
montanus Fall
pallidus Csy.
parallelus Csy.
pinalinus Csy.
provensus Csy.
pugetanus Csy.
punctatus Csy.
robustulus Csy.
shastanus Csy.
sonomæ Csy.
    LEUCORUS Csy. 05-191 [13]
ferrugineus Csy. (Leucorus)
luridus Csy. (Leucorus)
ochrinus Csy. (Leucorus)
rubens Csy. (Leucorus)
    PYCNORUS Csy. 05-194 [13]
dentiger Lec. (Pycnorus)
iowanus Csy. (Pycnorus)

Scopæus Er. 40-604 [12]
   (*Leptorus* Csy. 86-217) [13]
   (*Polyodontus* Sol. 49-310) [12]
   (*Pseudorus* Csy. 10-190) [13]
   (*Scoponæus* Mots. 58-641) [13]
    SCOPÆUS (s.str.) [12]
angustissimus Csy.
arizonæ Csy.
bicolor Csy.
brachypterus Csy.
carolinæ Csy.
cervicula Csy. (Pseudorus)
crassulus Csy.
degener Csy.

delicatus Csy.
exiguus Er.
gilensis Csy.
hudsonicus Csy.
longiceps Csy.
macilentus Csy.
notangulus Csy.
picipes Csy.
prolixipennis Csy. (Pseudorus)
quadripennis Csy.
saginellus Csy.
spectralis Csy. (Pseudorus)
texanus Csy.
versicolor Csy.
    SCOPÆCDERA Csy. 86-217 [13]
nitidus Lec. (Scopæodera)
sonoricus Csy. (Scopæodera)
    SCOPÆOPSIS Csy. 05-191 [13]
duryi Csy. (Scopæopsis)
elaboratus Csy. (Scopæopsis)
opacus Lec. (Scopæopsis)
pallens Csy. (Scopæopsis)
ventralis Csy. (Scopæopsis)
    SCOPÆOMA Csy. 05-191 [13]
angusticeps Csy. (Scopæoma)
caseyi Scheerp. 33-1265 [12]   Colo.
   procerus || Csy. 05-213 (Scopæo-
        [ma) [13]
notmani Scheerp. 33-1268 [12]
   pallidus || Notm. 21-152 (Scopæo-
        [ma) [13]
puritanus Csy. (Scopæoma)
rotundiceps Csy. (Scopæoma)
truncaticeps Csy. (Scopæoma)

Megastilicus Csy. 89-183 [13]
formicarius Csy.

Stilicolina Csy. 05-228 [13]
   (*Omostilicus* Csy. 05-229) [13]
sonorina Csy. (Omostilicus)
tristis Melsh.

Pachystilicus Csy. 05-226 [13]
hanhami Wickh.

Stilicus Latr. 25-495 [13]
   (*Rugilus* Curt. 27-168) [13]
   (*Stilicosoma* Csy. 05-219) [13]
abbreviellus Csy.
angularis Er.
apicalis Csy.
biarmatus Lec.
dentatus Say
lacustrinus Csy.
latiusculus Csy.
luculentus Csy.
minusculus Csy.
nigrolucens Csy.
opaculus Lec.
oregonus Csy.
rudis Lec.

Acrostilicus Hubb. 96-299 [13]
hospes Hubb.

[12] Scheerpeltz—33.
[13] Blackwelder—39.

Lathrobium Grav. 02-51 [13]
  (*Centrocnemis* Jos. 68-365) [13]
    LATHROBIUM (s.str.) [13]
      (*Litolathra* Csy. 05-71) [13]
amplipenne Csy.
amputans Csy. (Litolathra)
armatum Say
leconteanum Scheerp. 33-1278 [12]
  concolor || Lec. 63-44 (Litolathra) [12]
confusum Lec. (Litolathra)
convictor Csy. (Litolathra)
crurale Csy. (Litolathra)
dakotanum Csy. (Lathrobioma)
deceptivum Csy.
divisum Lec.
franciscanum Csy.
gravidulum Csy.
hesperum Csy. (Lathrobioma)
illini Csy.
innocens Csy.
inops Csy. (Lathrobioma)
inornatum Csy. (Litolathra)
longiventre Csy.
neglectum Csy.
nigrolineum Csy. (Lathrobioma)
nigrolucens Csy.
obtusum Csy.
oregonum Csy. (Lathrobioma)
othioides Lec. (Lathrobioma)
picescens Csy.
postremum Csy.
prælongum Csy.
procerum Csy.
rhodeanum Csy. (Litolathra)
rigidum Csy.
scolopaceum Csy. (Lathrobioma)
simile Lec.
simplex Lec.
sparsellum Csy.
spissicorne Csy.
subæquale Csy.
subgracile Csy. (Litolathra)
suspectum Csy. (Litolathra)
vancouveri Csy.
virginicum Csy. (Lathrobioma)
washingtoni Csy.
    LATHROLEPTA Csy. 05-72 [13]
debile Lec. (Lathrolepta)
    DERATOPEUS Csy. 05-73 [13]
nanulum Csy. (Lathrobioma)
nitidulum Lec. (Deratopeus)
parvipenne Csy. (Deratopeus)
semirubidum Csy. (Litolathra)
    TETARTOPEUS Czwal. 88-349 [13]
agitans Csy. (Tetartopeus)
angulare Lec. (Tetartopeus)
callidum Csy. (Tetartopeus)
captiosum Csy. (Tetartopeus)
finitimum Lec. (Tetartopeus)
floridanum Csy. (Tetartopeus)
furvulum Csy. (Tetartopeus)
hebes Csy. (Tetartopeus)
lacustre Csy. (Tetartopeus)
nigerum Lec. (Tetartopeus)
nigrescens Csy. (Tetartopeus)

punctulatum Lec. (Tetartopeus)
rubripenne Csy. (Tetartopeus)
semirubrum Csy. (Tetartopeus)
stibium Csy. (Tetartopeus)
terminatum Grav. (Tetartopeus)
tetricum Csy. (Tetartopeus)
    ABLETOBIUM Csy. 05-70 [13]
pallescens Csy. (Abletobium)
    APTERALIUM Csy. 05-70 [13]
brevipenne Lec. (Apteralium)
carolinæ Csy. (Apteralium)
    LATHROBIOPSIS Csy. 05-72 [13]
texana Csy. (Lathrobiopsis)
    LATHROBIOMA Csy. 05-72 [13]
tenue Lec. (Lathrobioma)

Lobrathium Muls. & Rey 78-29 [13]
  (*Bathrolium* Gozis 86-14) [13]
  (*Lathrobiella* Csy. 05-75) [13]
  (*Lathrotaxis* Csy. 05-74) [13]
    EULATHROBIUM Csy. 05-73 [13]
      (*Lathrotropis* Csy. 05-74) [13]
caseyi Blais. (Lathrotropis)
gnomum Csy. (Lathrotropis)
grande Lec. (Eulathrobium)
jacobinum Lec. (Lathrotropis)
puncticeps Lec. (Lathrotropis)
relictum Csy. (Lathrotropis)
subseriatum Lec. (Lathrotropis)
ustulatum Csy. (Lathrotropis)
vafrum Csy. (Lathrotropis)
validiceps Csy. (Lathrotropis)
    PSEUDOLATHRA Csy. 05-74 [13]
      (*Linolathra* Csy. 05-75) [13]
      (*Microlathra* Csy. 05-75) [13]
      (*Paralathra* Csy. 05-75) [13]
æmulum Csy. (Lathrobiella)
ambiguum Lec. (Lathrobiella)
anale Lec. (Pseudolathra)
angustulum Csy. (Lathrobiella)
angustum Csy. (Lathrotaxis)
atriventre Csy. (Lathrobiella)
bardum Csy. (Lathrobiella)
cupidum Csy. (Lathrobiella)
depressulum Csy. (Lathrobiella)
dimidiatum Say (Linolathra)
famelicum Csy. (Lathrobiella)
filicorne Csy. (Paralathra)
filitarse Csy. (Linolathra)
perfragile Scheerp. 33-1280 [12]
  fragile || Csy.05-139(Lathrobiella) [12]
gaudens Csy. (Linolathra)
gracilicorne Csy. (Lathrobiella)
habile Csy. (Lathrobiella)
inviolatum Scheerp. 33-1277 [12]
  integrum || Csy. 05-141 (Lathro-
      [biella) [12]
leviceps Csy. (Pseudolathra)
lineiforme Csy. (Microlathra)
lituarium Lec. (Linolathra)
merens Csy. (Lathrobiella)
modestum Csy. (Lathrobiella)
nigricans Csy. (Lathrobiella)
oregonense Csy. (Lathrobiella)

pallidulum Lec. (Microlathra)
robustulum Csy. (Lathrobiella)
rubidum Csy. (Lathrobiella)
rutilans Csy. (Microlathra)
tricolor Csy. (Lathrobiella)
vagans Csy. (Lathrobiella)
ventralis Lec. (Lathrobiella)
    LOBRATHIUM (s.str.)[13]
acomanum Csy. (Lathrotaxis)
atronitens Csy. (Lathrotaxis)
caseyianum Scheerp. 33-1274 [12]
   bipartitum || Csy. 05-122 [12]
californicum Lec. (Lathrotaxis)
canorum Csy. (Lathrotaxis)
centurio Csy. (Lathrotaxis)
collare Er. (Lathrobiella)
coloradense Csy.
expressum Csy. (Lathrotaxis)
fallaciosum Csy. (Lathrotaxis)
fallax Csy. (Lathrobiella)
floridæ Csy. (Lathrotaxis)
galvestonicum Csy. (Lathrotaxis)
longiusculum Grav. (Lathrotaxis)
montanicum Csy.
politum Grav. (Lathrotaxis)
præceps Csy. (Lathrotaxis)
rubricollis Csy. (Lathrotaxis)
soror Csy. (Lathrotaxis)
tacomæ Csy.

Acalophæna Shp. 86-554 [13]
   (*Calophæna* Lynch 84-267)[13]
compacta Csy.

Dacnochilus Lec. 63-47 [13]
lætus Lec.

Pæderus Fahr. 75-268 [13]
   (*Pæderomorphus* Gaut. 62-75)[13]
   PÆDERUS (s.str.)[13]
     (*Leucopæderus* Csy. 15-59)[13]
     (*Pæderidus* Muls. & Rey
          [78-245)[13]
     (*Pæderillus* Csy. 05-59)[13]
canonicus Csy.
carolinæ Csy.
compotens Lec.
femoralis Lec.
floridanus Aust.
grandis Aust.
iowensis Csy.
littorarius Grav.
nevadensis Aust.
obliteratus Lec.
pugetensis Lec.
riparius Linn.
saginatus Csy.
texanus Csy.
ustus Lec.
   NEOPÆDERUS Blkwr. 39-97 [13]
littoreus Aust.
palustris Aust.

Lissobiops Csy. 05-25 [13]
serpentinus Lec.

Homœotarsus Hochh. 51-34 [13]
   (*Spirosoma* Mots. 58-206)[13]
   GASTROLCBIUM Csy. 05-23 [13]
arizonensis Horn (Gastrolobium)
atriceps Csy. (Gastrolobium)
badius Grav. (Gastrolobium)
bicolor Grav. (Gastrolobium)
carolinus Er. (Gastrolobium)
coloradensis Csy. (Gastrolobium)
convergens Csy. (Gastrolobium)
despectus Lec. (Gastrolobium)
floridanus Lec. (Gastrolobium)
illinianis Csy. (Gastrolobium)
lecontei Horn (Gastrolobium)
melanocephalus Er. (Gastrolobium)
obliquus Lec. (Gastrolobium)
parallelus Csy. (Gastrolobium)
peninsularis Csy. (Gastrolobium)
pimerianus Lec. (Gastrolobium)
proximus Csy. (Gastrolobium)
spissiceps Csy. (Gastrolobium)
strenuus Csy. (Gastrolobium)
subatrus Csy. (Gastrolobium)
suturalis Csy. (Gastrolobium)
texanus Lec. (Gastrolobium)
vagus Horn (Gastrolobium)
ventralis Horn (Gastrolobium)
virginicus Csy. (Gastrolobium)
   HESPEROBIUM Csy. 05-33 [3]
atronitens Csy. (Hesperobium)
californicus Lec. (Hesperobium)
capito Csy. (Hesperobium)
cinctus Say (Hesperobium)
clavicornis Csy. (Hesperobium)
cribratus Lec. (Hesperobium)
flavicornis Lec. (Hesperobium)
pacificus Csy. (Hesperobium)
pallipes Grav. (Hesperobium)
parviceps Csy. (Hesperobium)
rubripennis Csy. (Hesperobium)
sellatus Csy. (Hesperobium)
tumidus Lec. (Hesperobium)
vancouveri Csy. (Hesperobium)

Cryptobium Mannh. 30-38 [13]
   ABABACTUS Shp. 85-533 [13]
pallidiceps Csy. (Ababactus)
   NEOBACTUS Blkwr. 39-96 [13]
nunenmacheri Blkwr. 39-96

Biocrypta Csy. 05-26 [13]
magnolia Blatch.
prospiciens Lec.

Stilicopsis Sachse 52-144 [13]
paradoxa Sachse
subtropica Csy.

[12] Scheerpeltz—33.
[13] Blackwelder—39.

Stamnoderus Shp. 86-607 [13]
  carolinæ Csy.
  monstrosus Lec.
  pallidus Csy.

Astenus Steph. 32-275 [13]
  (*Astenognathus* Reitt. 09-150)[13]
  (*Sunius* Er. 39-523, not Steph.)[13]
    ASTENUS (s.str.)[13]
  americanus Csy.
  arizonianus Csy.
  binotatus Say
  brevipennis Aust.
  californicus Aust.
  cinctus Say
  discopunctatus Say
  fusciceps Csy.
  inconstans Csy.
  linearis Er.
  longiusculus Mannh.
  ornatellus Csy.
  prolixus Er.
  robustulus Csy.
  sectator Csy.
  similis Aust.
  simulans Csy.
  specter Csy.
  strigilis Csy.
  tenuiventris Csy.
  zuni Csy.
  sinuaticollis Brg. 34-29   Pa.

Echiaster Er. 40-636 [13]
    ECHIASTER (s.str.)[13]
  ludovicianus Csy.
    LEPTOGENIUS Csy. 86-214 [12]
  brevicornis Csy. (Leptogenius)

Leptacinus [1]
  parumpunctatus Gyll. 27-481 [1] W.I. N.A.
      [Eur. Afr. Asia, Australia
    longicollis Steph. 33-259 [1]
    ampliventris Duval 54-37 [1]
    radiosus Peyron 58-421 [1]
    pallidipennis Mots. 58-206 [1]
    tricolor Kr. 59-110 [1]
    flavipennis Kr. 59-111 [1]
    amissus Fairm. & Coqu. 60-158 [1]
    breviceps Waterh. 77-24 [1]
    papuensis Fvl. 78-242 [1]
    sardous Fiori 94-94 [1]
    rubricollis Reitt. 99-157 [1]
    fauveli Cam. 22-114 [1]

[1] Blackwelder in MS.

STAPHYLININI

Diochus Er. 39-300
  • nanus Er. 39-301 [1] S.A. C.A. W.I. N.A.
    schaumi Kr. 60-27 [1]
    parvulus Kr. 60-27 [1]
    longicornis Shp. 76-184 [1]
    vicinus Shp. 76-185 [1]
    tarsalis Shp. 76-185 [1]
    flavicans Shp. 76-185 [1]
    inornatus Shp. 85-466 [1]
    vilis Shp. 85-467 [1]
    maculicollis Fvl. 91-106 [1]
    brevipennis Csy. 06-431 [1]
    thoracicus Csy. 06-432 [1]
    pallidiceps Csy. 06-432 [1]
    perplexus Cam. 22-116 [1]
    apicipennis Cam. 22-116 [1]
    antennalis Cam. 22-117 [1]
    pumilio Bnhr. 29-193 [1]

Philonthus
  lacustrinus Scheerp. 33-1347 [12]
    lacustris || Csy. 15-432 [12]
  thermarum Aube 50-316 [1] S.A. W.I. Mo.
      [D.C. Eur. Asia, Afr.

    exilis Kr. 51-292 [1]
    angustatus Kr. 59-92 [1]
    pygmæus Kr. 59-93 [1]
    fuscolaterus Mots. 59-76 [1]
    inclinans Walker 59-51 [1]
  flavolimbatus Er. 40-471 [1]   S.A. C.A.
                          [W.I. Fla.-Cal.

    apicipennis Lynch 84-155 [1]
  hepaticus Er. 40-451 [1]   S.A. C.A. W.I.
    [Cal. Nev. Ariz. Pa. Kan. Australia
    vilis Er. 40-451 [1]
    orphanus Er. 40-452 [1]
    nanus Melsh. 46-36 [1]
    cinctutus Melsh. 46-37 [1]
    palleolus Melsh. 46-37 [1]
    rufipennis Sol. 49-317 [1]
    varicolor Boh. 58-29 [1]
    pyropterus Kr. 59-12 [1]
    pauxillus Solsky 67-133 [1]
    parvimanus Shp. 85-406 [1]
    cinctulus Bnhr. & Schub. 14-341 [1]
  varians Payk. 89-45 [1]   W.I. N.A. Eur.
                          [Afr. Asia.

    bipustulatus Grav. 02-37 [1]
    aterrimus Marsh. 02-513 [1]
    nitens Grav. 02-26 [1]
    opacus Grav. 02-26 [1]
    bimaculatus Marsh. 02-525 [1]
    intaminatus Steph. 32-235 [1]
    lituratus Steph. 32-238 [1]
    punctiventris Steph. 32-235 [1]
    unicolor Steph. 32-224 [1]
    costatus Baudi 48-29 [1]
    incompletus Hochh. 49-153 [1]
    scutatus Epp. 95-127 [1]
    alpigradus Muls. & Rey 75-481 [1]
    proteus Everts 22-124 [1]
    brunneipennis Everts 22-124 [1]
    piceicornis Grid. 20-18 [1]
    fuscicoxis Scheerp. 33-1366 [1]

longicornis Steph. 32-237 [1]   S.A. C.A.
    [W.I. N.A. Eur. Afr. Asia, etc.
fuscicornis Nord. 37-96 [1]
scybalarius Nord. 37-94 [1]
feralis Er. 40-469 [1]
promtus Er. 40-929 [1]
varians Fairm. 49-290 (not Payk.)[1]
algiricus Mots. 58-663 [1]
pedestris Walk. 59-51 [1]
asemus Kr. 59-86 [1]
perplexus Fairm. & Germ. 61-431 [1]
fumosus Sols. 68-134 [1]
lætabilis Olliff 87-501 [1]
linkei Bnhr. 08-34 [1]
rubromaculatus Bnhr. 15-9 [1]
piceicornis Grid. 20-18 [1]
ventralis Grav. 02-174 [1] S.A. W.I. N.A.
    .   [Eur. Asia, Afr.
anthrax Grav. 02-176 [1]
immundus Grav. 06-66 [1]
celer Grav. 06-66 [1]
picicollis Steph. 32-224 [1]
fulvipes Steph. 32-229 (not Fabr.)[1]
rotundiceps Steph. 32-248 [1]
proximus Woll. 57-189 [1]
fortunatus Woll. 65-493 [1]
discoideus Grav. 02-38 [1] W.I. N.A. Eur.
    [Afr. Asia, Hawaii
testaceus Payk. 89-28 [1]
suturalis Marsh. 02-509 [1]
lepidulus Steph. 32-223 [1]
conformis Bois. & Lac. 35-398 [1]
ruficornis Melsh. 46-38 [1]
rufipennis Gerb. 10-555 [1]
gerhardtianus Scheerp. 33-1340 [1]
kessleri Blatch. 36-256
multipunctatus || Blatch. 10-389

Belonuchus
rufipennis Fahr. 01-597 [1] S.A. C.A. W.I.
formosus Grav. 06-72 [1]   [N.A.
apicalis Dej. 33-63 [1]
pallipes Melsh. 46-35 [1]
schæfferi Cooper 33-545   Tex.

Cafius Steph. 32-345 [14]
*   CAFIUS (s.str.) [14]
nudus Shp. 74-36 [14]   Japan, Vanc.
johnsoni Fall 16-13 [14]
    BRYONOMUS Csy. 86-313 [4]
canescens Makl. 52-313 [14]   Cal.-Alas.
seminitens Horn 84-236 [14]   Cal.
    EUREMUS Brg. 34-68 [14]
decipiens Lec. 63-40 [14]   Cal.
sulcicollis Lec. 63-40 [14]   Cal.
luteipennis Horn 84-237 [14] L.Cal.-Vanc.
bistriatus Er. 40-502 [14] N.Y.-Fla. W.I.
bilineatus Er. 40-503 [14]
rufifrons Brg. 34-68 [14]
    PSEUDOREMUS Koch 36-179 [14]
opacus Lec. 63-40 [14]   Cal.
lithocharinus Lec. 63-38 [14]   Cal.
    REMUS Holme 37-64 [14]
sericeus Holme 37-64 [14]   Eur. Atl.Cst.
f.aguayoi Brg. 34-66 [14] [15]   Atl.Cst.
f.femoralis Mäkl. 53-189 [14] Alas.-Cal.

Staphylinus
caseyi Scheerp. 33-1392
quadraticeps || Csy. 25-149 [12]
macgregori Cooper 33-264   Ariz.
globulifer Fourc. 85-164 [16]   Eur. Que.
edentulus Block 99-115 [16]
morio Grav. 02-6 [16]

Creophilus
* maxillosus Linn. 58-421 [1] C.A. W.I. N.A.
                        [Eur. Asia, Afr.
anonymus Sulz. 61-17 [1]
balteatus DeG. 74-18 [1]
fasciatus Fuess. 75-21 [1]
nebulosus Fourcr. 85-165 [1]
villosus Grav. 02-160 [1]
ciliaris Steph. 32-202 [1]
arcticus Er. 39-350 [1]
cinerarius Er. 39-350 [1]
bicinctus Mannh. 43-229 [1]
orientalis Mannh. 43-229 [1]
fulvago Mots. 60-120 [1]
imbecillus Shp. 74-28 [1]
medialis Shp. 74-28 [1]
subfasciatus Shp. 74-28 [1]
pulchellus Meier 99-99 [1]
canariensis Bnhr. 08-334 [1]

Quedius
caseyi Scheerp. 33-1435   Man.
curtipennis || Csy. 15-414 [12]

Anacyptus Horn 77-87
(Microcyptus Horn 82-1)[1]
* testaceus Lec. 63-30 [1] W.I. Fla.-Ariz.-
                        [L.Sup.

Tachyporus Grav. 02-124 [17]
* nitidulus Fahr. 81-337 [17] Mass.-Va.-Cal.-
brunneus Fahr. 92-535 [17]   [B.C.
faber Say 34-468 [17]
scitulus of Horn 77-105 [17]
nanus Er. 40-240 [17]   D.C. Pa. Mich.
rulomus Blkwr. 36-44 Alas.-Cal.-Mich.
tehamæ Blkwr. 36-45   Cal.
californicus Horn 77-104 [17] Cal.-Wash.
stejnegeri Blkwr. 36-47   Alaska
jocosus of Lin. & Schw. 98-333 [11]
maculipennis Lec. 66-374 [17]   Vt.-Ga.-
                        [N.M.-B.C.
elegans Horn 77-103 [17]   Mass.-W.Va.-
                        [Ckla.-Man.
snyderi Blkwr. 36-49   D.C.-Fla.-Ky.
temacus Blkwr. 36-49   Colo. Nev. Sask.
jocosus Say 34-466 [17]   N.H.-Fla.-Cal.-
arduus Er. 40-237 [17]   [Y.T.
pulchrus Blatch. 10-447 [17]   Ind.
oregonus Blkwr. 36-51   Ore. Cal.
acaudus Say 34-467 [17] Mass.-Ky.-N.M.-
                        [Wash.-Ont.
maculicollis Lec. 66-374 [17]
heterocerus Lec. MS [17]
angusticollis Lec. MS [17]
arizonicus Blkwr. 36-53   Ariz.
alleni Blkwr. 36-53   Ore.
mexicanus Shp. 83-311 [17]   Mex. Tex.

[1] Blackwelder in Ms.
[14] Revision of genus. Koch—38.
[15] Described as valid species. Bierig—34.

[12] Scheerpeltz—33.
[16] Chagnon—36.
[17] Revision of genus. Blackwelder—36.

Leucoparyphus Kr. 57-393
* silphoides Linn. 67-684 [1] W.I. N.A. Eur.
[Asia, Afr.
  suturalis Panz. 94-18 [1]
  marginalis Grav. 02-192 [1]
  limbatus Grav. 06-12 [1]
  pictus Er. 39-246 [1]
  geminatus Rand. 38-39 [1]

Coproporus Kr. 57-399 [18]
* (*Erchomus* Mots. 58-218)[18]
  (*Cilea* Pand. 69-277, not Duval)[18]
  ventriculus Say 34-466 [18] N.H.-Fla.-Cal.-
[Wash.-Man.
  acuductus Kby. 37-90 [18]
  affinis || Kby. 37-91 [18]
  gibbulus Er. 39-252 [18]
  punctulatus Melsh. 46-32 [18]
  flavidus Csy. 84-141 [8]
  politus || Manee 15-175 [18]
  maneei Scheerp. 34-1517
rutilus Er. 39-253 [18] S.A. C.A. W.I. Tex.
  brevis Scriba 55-296 (not Sharp)[18]
lecontei Blkwr. 38-5   Ariz. Cal.
  punctipennis || Lec. 63-31 [18]
lævis Lec. 63-31 [18]   Md.-Fla.-Tex.-Ind.
sparsus Blkwr. 38-7   Ariz.
hepaticus Er. 39-249 [1]   S.A. C.A. W.I.
  convexus Er. 39-248 [1]   [Tex.-Cal.
  ignavus Shp. 76-87 [1]
  inflatus Horn 77-107 [1] [18]
arizonæ Blkwr. 38-8   Mex. Ariz. Tex.
pulchellus Er. 39-247 [1]   S.A. C.A. W.I.
  cumanensis Scriba 55-297 [1]   [Fla.
  infimus Duval 57-33 [18]
  distans Shp. 76-92 [1]

Philotermes Kr. 57-13 [19]
  pilosus Kr. 57-14 [19]   Mass.-D.C.-Ill.
  fuchsii Kr. 57-15 [19]   Fla.-Tenn.-Ark.
  pennsylvanicus Kr. 57-15 [19] Mass.-D.C.-
[Ill.-Tex.
  emersoni Seevers 38-435 [19]   Ind.

Atheta
  fenyesiana Scheerp. 34-1607
  annuliventris || Fenyes 14-49 [12]
  occidentalis Bnhr. (Cat. No. 5108 &
[No. 5133)

Baryodma
  bilineata Gyll. 10-436 [4]   Eur. N.A.
Termitonidia Seevers 38-428 [20]
  lunata Seevers 38-429   Ariz.

Eburniogaster Seevers 38-424 [20]
  termitocolus Seevers 38-426   Ariz.
  texanus Brues 02-186 [21]   Tex.

[4] Voris in litt.
[18] Revision of genus, Blackwelder—38.
[19] Revision of genus, Seevers—38.
[20] Should follow Termitogaster in Aleocharini
(near Ocalea).
[21]Seevers—38.

PSELAPHIDÆ

Actium
  pennsylvanicum Bowm. 34-141   Pa.

Euplectus
  excavatus Bowm. 34-141   Pa.

Trigonoplectus Bowm. 34-37 [1]
  minutus Bowm. 34-38   Pa.
  rostratus Bowm. 34-38   Pa.

Bibloplectus
  exilis Bowm. 34-141   Pa.

Batrisodes
  kahli Bowm. 34-141   Tenn.

Brachygluta
  mormon Bowm. 34-142   Utah

Reichenbachia
  utahensis Tanner 34-43   Utah

Decarthron
  setosum Bowm. 34-142   Pa.

Pselaphus
  ulkei Bowm. 34-142   Can. Mass. Ida.
[S.D.

HISTERIDÆ [2]

Hololeptinæ [2]

Hololepta [2]
Iliotona [3]

Teretriinæ [2]

Teretrius [2]
Teretriosoma [2]

Abræinæ [2]

Peploglyptus [2]
Onthophilus [2]
Plegaderus [2]
  setulosus Ross 38-50   B.C.
Bacanius [2]

[1] To precede Acalonia, Bowman—34.
[2] An alternative arrangement of genera differing
from Leng's, is given by Bickhardt in Genera
Insectorum fasc. 166, and above.
[3] Leng Catalog.

Anapleus [2]
  marginatus Lec. 53-292
    compactus Csy. 93-558 [4]
    mexicanus Csy. 16-248 [4]

Abræus [2]

Acritus [2]

Saprininæ [2]

Saprinus [2]
  felipæ Lewis 13-87   Tex.[5]
  alienus Lec. 51-167
    shantzi Csy. 24-200 [4]
  pennsylvanicus Payk. 11-62
    profusus Csy. 93-566 [4]
  semistriatus Scriba 90-76   Eur. Afr.
                              [Mex. Ia.-Mich.
    semipunctatus Payk. 98-45 [6]
    acuminatus Fabr. 98-37 [9]
    nitidulus Fahr. 01-85 [6]
    incrassatus Men. 32-170 [6]
    krynickii Kryn. 32-113 [4]
    turcomanicus Men. 48-55 [6]
    subattenuatus Mots. 49-95 [6]
    planiusculus Mots. 49-97 [4]
    sparsipunctatus Mots. 49-97 [6]
    uralensis Mots. 49-98 [4]
    punctostriatus Mars. 62-460 [6]
    steppensis Mars. 62-460 [4]
    rugipennis Hockh. 72-225 [6]
    hockhuthi Reitt. 06-267 [6]
    subnitescens Bickh. 09-221 [6]
    imperfectus of Blatch. 10-620 [6]
    lecontei Csy. 16-262 [6]
    pacoviensis Roubal 27-94 [4]
  assimilis Payk. 11-63
    simulatus Blatch. 10-621 [6]
    conformis of Blatch. 10-620 [6]
    semisulcus Hatch 29-79 [7]
  conformis Lec. 45-72
    oviformis Blatch. 10-622 [4]
  sphæroides Lec. 45-77
    impunctellus Csy. 93-571 [6]
    lakensis Blatch. 10-623 [6]
    illinoensis Wolc. 12-161 [6]
    eriensis Hatch 29-82 [6]
    ontarioensis Hatch 29-82 [6]
    ohioensis Hatch 29-82 [6]
    michiganensis Hatch 29-83 [4]
    seminitens of Blatch. 10-623 [4]
  alutiger Wenzel 35-189   Wis. Ind.
  prosternalis Hinton 35-78   L.Cal.
  strigithorax Hinton 35-79   L.Cal.
  ferrugineus Mars. 55-712
    lustrans Csy. 16-269 [4]
  bigemmeus Lec. 51-169
    parvus Csy. 16-273 [4]
    strigilarius Csy. 16-274 [4]
  fitschi Mars. 62-494
    omissus Csy. 16-272 [4]
    ludovicianus Csy. 24-206 [4]

[2] An alternative arrangement of genera differing
from Leng's. is given by Bickhardt in Genera-
Insectorum fasc. 166. and above.
[4] Wenzel in litt.
[5] Ballou in litt.
[6] Wenzel—39.
[7] Wenzel—35.

lucidulus Lec. 51-170
  recticollis Csy. 24-204 [4]
  diego Csy. 24-204 [4]
  obsolescens Csy. 24-205 [4]
  dimidiatipennis Lec. 24-170
  palmatus Say 25-42 (or aberr.)[8]

Bæckmanniolus Rchdt. 26-12 [8]
  balloui Hntn. 35-80 [9]   L.Cal.
  gaudens Lec. 51-165   Cal. Mex.
    guardens Hntn. 35-81 [9]
  palmatus Say 25-42 [9]   E.U.S.
  serrulatus Lec. 51-165 [9]   Cal.

Chelyoxenus [2]
  * xerobatis Hubb. 94-309   Fla.
    repens Csy. 16-262 [4]
    insolitus Csy. 16-263 [4]

Gnathoncus [2]

Dendrophilinæ [2]

Dendrophilus [2]
  punctatus Hbst. 92-41
    punctulatus Say 25-45 [10]
    xavieri Lewis [10]   Japan
  sexstriatus Hatch 38-18   Iowa
  tularensis Ross 37-67   Cal.

Xestipyge [2]

Carcinops [2]
  gilensis Lec. 51-164
    uteana Csy. 16-242 [4]
  consors Lec. 51-164
    papagoana Csy. 93-554 [4]
    nigra Csy. 16-242 [4]
  opuntiæ Lec. 51-164
    bisculpta Csy. 16-243 [4]
    perlata Csy. 16-244 [4]

Paromalus [2]

Isolomalus [2]
  seeversi Wenzel 37-266   Kan.
  seminulum Er. 34-171
    ovulatus Csy. 16-246 [4]

Histerinæ
TRIBALINI [8]

Epierus [2]

Stictostix [2]

Tribalister [2]

Idolia [2]

Cærosternus [2]

[8] Leng & Mutchler—34.
[9] Hinton—35.
[10] Wenzel in litt. (after Ross).

PLATYSOMINI [2]

Cylistix [2]
cylindrica Payk. 11-91
parvula Csy. 24-198 [4]

Platysoma [2]
depressum Lec. 45-40
pinorum Csy. 16-202 [4]
tabellum Csy. 93-551 [4]

Omalodes [2]

HISTERINI [2]

Psiloscelis [2]

Hister [2]
felipæ Lewis 01-373 [5]   Kans.
ænigmaticus Wenzel 37-267   Ind.
semiruber Csy. 93-539
solaris Carn. 15-144 [4]
militaris Horn 70-135   Cal. Ore. Wash.
?oregonus Csy. 93-549 [11]
?electus Csy. 93-548 [11]
?simplicipes Fall 01-235 [11]
ciliatus Lewis 88-199 [2]   Mex. Ariz.
sexstriatus Lec. 51-163
maritimus Csy. 16-214 [11]
s.jacobianus Csy. 16-215 [11]
stygicus Lec. 45-48
jaquesi Hatch 29-76 [6]
interruptus Beauv. 05-180
immunis Er. 34-143 [6]
albertensis Hatch 26-275 [6]
carri Hatch 26-276 [6]
fœdatus Lec. 45-50
texensis Csy. 24-217 [4]
PARALISTER Bickh. 16-188
cognatus Lec. 45-28 [13]   N.Y.-Can.-Tex.
unicus Csy. 93-547 [6][13]
sinuosus Lewis 00-231 [13]
marginicollis Lec. 45-28 [13]   Ill. Ind.
remotus Lec. 59-70 [13]   Cal. Ore.
semisculptus Lec. 63-60 [13]   Ill.
californicus Mars. 54-115 [13] L.Cal. Mex.

* * *

abbreviatus Fab. 75-53
coloradensis Csy. 24-198 [4]
depurator Say 25-33
furtivus Lec. 59-313 [4]
circinans Csy. 16-220 [4]
osculatus Blatch. 10-607
puncticollis Schffr. 12-26 [6]
grandis Wenzel 39-13   Iowa
fungicola Schffr. 12-27
nanulus Csy. 16-224 [4]
indistinctus Say 25-35
debilicinus Csy. 24-197 [4]
densicauda Csy. 16-222
cribricauda Csy. 16-222 [4]
americanus Payk. 11-31
diffractus Csy. 16-225 [4]

[11] Ross—37.
[12] Ballou & Siepmann—88.
[13] Revision of subgenus, Wenzel—37.

pollutus Lec. 59-7
lævicauda Csy. 16-227 [4]
fluviatilis Csy. 16-228 [4]
tornatus Lec. 80-190 [4]   N.C. Ga.

Margarinotus [2]

EXOSTERNINI [2]

Phelister [2]
(*conquisitus* Lewis, not No. American) [9]

Pseudister [9]
hospes Lewis 02-236 [3]   N.Y.

Hetæriinæ [2]

HETÆRIOMORPHINI [2]

Terapus [2]
arizonensis Ross 38-48   Ariz.

Ulkeus [2]

HETÆRIINI [2]

Reninus [4]
salvini Lewis 88-220 [4]   Tex.

Echinodes [2]

Hetærius [2]
wagneri Ross 38-49   Cal.

LYCIDÆ

Plateros
californicus VanD. 18-1
columbiensis Brown 29-108 [1]

LAMPYRIDÆ

Aspidosoma
ignitum Linn. 67-645 [2]   Tex. Mex.-Bra.

Phausis
nigra Hopp. 37-89   B.C.

CANTHARIDÆ

Cantharis
perpallens Fall 36-179   Cal.
imbecillus Lec. 51-342   Pa. Conn.
mimus Fall 36-182   N.J.
mollis Fall 36-182   N.H.-Pa.
pusillus of Lec. 1881 (not Lec. 1851) [3]

[1] Fall—34.
[2] Fall—37.
[3] Fall—36.

greenei Fall 36-183   W.Va.   **Allonyx** [5]
lecontei Fall 36-180
   collaris || Lec. 51-340 [3]   **Pseudallonyx** [5]
tuberculatus Lec. 51-341
   impressus Lec. 51-341 [3]   **Vectura** [5]
   brevicollis Lec. 51-341 [3]   fulvescens Blais. 34-71   Cal.
   armiger Couper 65-62 [3]

                             Dasytastes [5]

MELYRIDÆ [4]   **Mecomycter** [6]

Malachinæ [4]   **Pseudasydates** Blais. 38-18 [5]
                                         inyoensis Blais. 38-18   Cal.
Collops
  bridgeri Tanner 36-153   Wyo.   **Eudasytes** [5]
                                         sinuatus Blais. 37-138   Cal.
Malachius
  utahensis Tanner 36-153   Ut. Ida. **Asydates** [5]
  antennatus Hopp. 37-90   B.C.
                                       Sydatopsis [5]
Tanaops
  testaceus Marshall 37-164   Ariz.   **Cradytes** [5]

Attalus   **Eutricholistra** [5]
  smithi Hopp. 37-91   B.C.
                                       Byturosoma [5]

Rhadalinæ [4]   **Trichochrous** [5]
                                         cupripilosa Blais. 37-140   Cal.
                                         calcaratus Fall 34-142   Is. off Cal.
Melyrinæ [4]
                                       Trichochronellus Blais. 38-23 [5]
DASYTINI [5]   stricticollis Csy. 95-532 [5][10]   Cal.

Cymbolus [5]   **Holomallus** Gorh. 86-325 [5]

Eucymbolus [5]   **Eutrichopleurus** Blais. 38-24 [5]
                                         seriellus Csy. 95-506 [5][10]   Utah
Dolichosoma [5]
                                       Emmenotarsus [5]
Pristoscelis [5]
                                       Trichochroides Blais. 38-25 [5]
Eschatocrepis [5]   sexualis Csy. 95-524 [5][10]   Cal.

Dasytes [5]   **Sydates** [5]
  blaisdelli Pic 37-65   U.S.
  subœneus Blais. 26-12 [4]   **Listromimus** [5]

Dasytellus [5]   **Listromorpha** Blais. 21-188 [5]

Vecturoides Fall 30-255 [5]   **Adasytes** [5]
  (*Menovectura* Blais. 31-181) [5]
                                       Amecocerus Sol. 49-419 [4]
Amphivectura Blais. 38-13 [5]   (*Listrus* Mots. 59-389) [4][11]
  monticola Blais. 34-151 [5][7]   B.C.   caseyi Pic 37-99   U.S.
                                         subœneus || Csy. 95-342 [4]
Hoppingiana Blais. 24-2 [5]   regalis Blais. 38-165   Cal.
  hudsonica Lec. 66-360   vanduzeei Blais. 34-317   Cal.
  brevilabris Blais. 24-3 [5][9]   coalingensis Blais. 36-187   Cal.
  kingi Brown 28-147 [9]   gentryi Blais. 36-185   Cal.
                                         minimus Blais. 36-184   Cal.
Leptovectura [5]   coronadensis Blais. 39-57   L.Cal.
                                         robustus Blais. 37-143   Wyo.
                                         wyomingensis Blais. 37-141   Wyo.

[3] Fall—36.   **Listropsis** Blais. 24-1 [5]
[4] Subfamilies rearranged by Blaisdell—38.
[5] Genera rearranged by Blaisdell—38.
[6] Pic—37.   *MELYRINI* [5]
[7] Described in Hoppingiana, Blaisdell—34.
[8] Blaisdell—34.   [10] Described in Trichochrous.
[9] Blaisdell—38.   [11] Listed as valid by Blaisdell—38.

CLERIDÆ

The proper date of the following Leconte species should be 1849 instead
of 1852: 7531, 7545 syn., 7561, 7565,
7597 syn., 7602 syn., 7610 syn., 7614,
7615, 7630a syn., 7643, 7647a, 7647a
syn., 7660, 7661, 7679, 7681, 7689, 7714,
and the genus Elasmocerus.[12]

Tillinæ [12]

Monophylla Spin. 41-75 [13]
(Elasmocerus Lec. 52-13)[13]
(Macrotelus Spin. 41-75)[13]

Callotillus Wolc. 11-115 [13]
vafer Wolc. 21-270   Ariz. S.Cal[2]

Tillus Cliv. 90-22 [13]
(Cylinder Voet 69-06—78)[13]
(Tilloidea Cast. 32-398)[13]

Perilypus Spin. 41-72 [13]

Cymatodera Gray 32-375 [13]
neomexicana Knull 34-9   N.Mex.
xanti Horn (Locality should be L.Cal.)[12]
peninsularis Schffr. (Locality should be
[Lower Cal., Ariz.)[12]
fuchsi Schffr. 04-216
?comans Wolc. 10-351 [12]
7563, for 65-95 read 66-95 [12]
7568a, for 65-95 read 66-95 [12]
7569, for 04-127 read 04-217 [13]
cephalica Schffr. (Locality should be
[Cal.)[12]
purpuricollis Horn (Locality should be
[Lower Cal.)[12]

Bostrichoclerus VanD. 38-189
bicornis VanD. 38-190   Gulf of Cal. Is.

Hydnocerinæ [13]

Hydnocera Newm. 38-379 [13]
7640, for 08-333 read 08-133 [12]
7642, for 65-97 read 66-97 [12]
7668, for 67-135 read 68-135 [12]
(lateralis Gorh., not No. American)[12]
suturalis Klug
limbata Spin. 44-49 [12]
verticalis Say
brachyptera Klug 42-313 [12]

Isohydnocera Chpn. 17-83 [13]
nigrina Schffr. 08-134 [12]   Ariz.
7687, for 65-97 read 66-97 [12]

(Erolestes Wolc., not North American)[12]
(cleroides Wolc., not No. American)[12]

[12] Wolcott in litt.
[13] New generic classification after Chapin—24 and
Schenkling—16. with synonymic additions from
Schenkling—Col. Cat. p. 23.

Phyllobæninæ [13]

Phyllobænus Spin. 44-1 [13]

Clerinæ [13]

Priocera Kby. 18-389 [13]
chiricahuæ Knull 39-27   Ariz.

Opilo Latr. 02-111 [13]
(Eupocus Ill. 07-341) [13]
(Notoxus of authors)[13]
(Opilus of authors)[13]

Serriger Spin. 41-73 [13]
Ref. to Wolcott on p. 28 Suppl. 1 should
[read 22-77.[12]

Cleronomus Klug 42-282 [13]
(Derestenus Chev. 43-13)[13]
(Colyphus Spin. 44-133)[13]
furcatus Schffr. 04-218 [12]   Tex.
haagi Chevr. 76-12 [12]   Mex. Tex.
melanopterus Dury 06-251 [12]   Ohio
subcostatus Schffr. 17-131 [12]   Fla.
thoracica Cliv. 95-18 [12]
monilis Melsh. 46-307 [12]
ornaticollis Lec. 80-194 [12]
v.pallipes Wolc. 12-55 [12] Ill. Kan. Ia.
[Neb.

Thanasimus Latr. 06-270 [13]
(Cleroides Schäff. 77-137)[13]
(Pseudoclerus Duval 60-196)[12]
7582, for 70-342 read 71-342 [12]
undatulus Say 35-163
undulatus of authors [12]

Clerus Fahr. 75-157 [13]
(Enoclerus Gahan 10-62)[13]
lecontei Wolc. 10-359
nigriventris || Lec. 61-351 [12]
liljebladi Wolc. 22-73 Mich. N.Y.[12] Pa.[12]
erro Wolc. 22-68 (Thanasimus)[12] Ariz.
inyoensis VanD. 38-191   Cal.
(quadriguttatus Oliv., not No. Amer.)[12]
abdominalis Chevr. 34-52   Mex. Ariz.
zonatus Klug 42-297 [12]
v. spinolæ Lec. 53-230 [12] Mx. Ar. Cal.
[Kan.[12]
bombycinus Chevr. 33-42   Cal.[12]
sphegeus Fabr. 87-125
sobrius Walk. 66-326
sobrinus of authors (not Cast.) [12]

Placopterus Wolc. 10-363 [13]
(Phlœpterus of authors)
(Plœopterus of authors)
(Pœcilochroa Chevr. 76-5) [13]
cyanipennis Klug 42-307 [14]   Mex.
v.dasytoides Westw. 49-50 [14] Mex. Tex.

[14] Only representative in North America—Wolcott—in litt.

Aulicus Spin. 41-74 [13]
    (*nero* Spin., *not North American*)[12]
    humeralis Linsley 36-252 [15]   L.Cal.
                    [Gulf Cal.Is.
    femoralis Schffr. 17-132 [15]   Ariz. Tex.
    monticola Gorh. 82-146 [15] Mx. Tex. Ar.
    fissipes Schffr. 21-155 [15]   Ariz. L.Cal.
    nigriventris Schffr. 21-156 [15] Mex. Ariz.
    dentipes Schffr. 21-157 [15]   Tex. Ariz.
    terrestris Linsley 33-95 [15] [16]   Cal.
    bicinctus Linsley 36-255 [15]   Cal.

(Sallœa Chevr., not North American) [11]
(coffini White, not North American) [12]

Xenoclerus Schenk. 02-327 [13]

Trichodes Hbst. 92-154 [13]
    (Pachyscelis Hope 40-139) [13]
    interruptus Lec., for 52-88 read 49-18 [12]
    peninsularis Horn (Locality should be
                    [Lower Cal.) [12]
    ornatus Say (Locality should be Dak.
               [Colo.-Pac.-Alas.) [12]
    nutalli Kby. (Add locality Colo.) [12]

Thaneroclerinæ [11]

Thaneroclerus Lef. 38-13 [13]
    (Isoclerus Lewis 92-191) [13]
    (Thaneclerus Chenu 60-247) [12]
    (Thanateroclerus Gemm. & Har.
                   [69-1739) [13]

Ababa Csy. 97-653 [12]
    (Prionostichœus Wolc. 11-125)
    (Prionodera || Wolc. 10-396)

Enopliinæ [13]

Ichnea Cast. 36-55

Chariessa Perty 30-109 [13]
    (Brachymorphus Chevr. 35-150) [13]
    7707, for 57-48 read 60-48 [12]

Pelonium Spin. 44-347 [13] [17]
    (Corinthiscus Frm. & Grm. 61-4 [13]
    (Philyra Cast. 36-53) [13]
    (Tarandocerus Chevr. 76-7) [13] [17]

Oregya Lec., [13] for 62-197 read 61-197 [12]
    7717, for 65-98 read 66-98 [13]

Orthopleura Spin. 44-80 [13]
    (Dermestoides Schaeff. 71-220) [13]
    damicollis
        bimaculata Melsh. 46-307 [12]
    (quadraticollis Spin., not N. Amer.)[12]

Pelonides Kuw. 94-8 [13]
    (Enoplium of authors) [13]
    7699, for 67-135 read 68-135 [12]
    (militaris Chevr., not No. American) [11]

[12] Wolcott in litt.
[13] New generic classification after Chapin--24 and Schenkling—16, with synonymic additions from Schenkling—Col. Cat. p. 23
[15] Revision of genus, Linsley—36.
[16] Linsley—33-88.
[17] Listed as synonyms of Chariessa by Wolcott in litt.

similis Knull 38-97   **Tex.**
perroudi Pic 33-12 (Pyticera)   **Tex.**
7704, for 65-99 read 66-99 [12]

Korynetinæ [13]

Lebasiella Spin. 44-77 [13]
    marginella Chevr. 43-42 [12] ?Cal. Nev.

Korynetes Hbst. 92-148 [13]
    (Corynetops Duval 61-201) [13]
    (Corynetes of authors) [13]
    maculicollis
      v.nigricollis Wolc. 27-110 [12]   Cal.

Necrobia Cliv. 95-76 [13]
    (Agonolia Muls. 64-122) [13]

Tarsostenus Spin. 44-287 [13]

Opetiopalpus Spin. 44-110 [13]
    (Opetiopselaphus Gemm. & Har.
                 [69-1759) [13]

CEPHALOIDÆ

Cephaloon Newm. 38-377 [18]
    ungulare Lec. 74-275 N.H.-N.C. L.Sup.
    pacificum VanD. 28-260   Cal.-B.C.
    lepturides Newm. 38-377 Me.-Pa.-Mich.
      lepturoides Hald. 48-95 [18]
      varians Hald. 48-95 [18]
    bicolor Horn 96-381   Cal. B.C.
    tenuicorne Lec. 74-275 B.C.-Alb.-Wash.-
      piceum Horn 96-380 [18]   [Mont.
      ornatum Csy. 97-652 [18]
      versicolor Csy. 97-651 [18]
    vandykei Hopp. & Hopp. 34-69   Cal.

MORDELLIDÆ

Mordellistena [19]
    frosti Lilj.   **Me.**
    erratica Smith   **Fla.**
    bifasciata Ray 36-125   **Ill.**
    bicinctella Lec.   Fla. S.St. Ind. O.
    confusa Blatch.   **Ind.**
    ozarkensis Ray 36-125   **Ill.**
    tarsalis Smith   **Tex.**
                 • • •
    rubrofrontalis Ray 36-127   **Ind.**
    tiara Ray 36-127   Ill. Mass.
    rufocephala Ray 36-128   **Ill.**

Pentaria Muls. 56-391 [20]
    (Anthobates || Lec. 50-23) [20]
    (Anthobatula Strand 29-23)[20]

[18] Revision of genus, Hopping & Hopping—34.
[19] Revision of part of section I, Ray—36.
[20] Mequignon—37.

MELOIDÆ [21]

Nemognathinæ

ZONITINI

Zonitis Fahr. 75-126 [21]
  (Nemognatha Ill. 07-333) [21]
  (Nematognatha Gemm. & Har.
    [70-2163) [21]
apicalis Lec. 53-345   B.C. Cal. Ariz.
  bicolor Walk. 66-331 [21]
  walkeri Beaur. 89-212 [21]
arizonica VanD. 29-132
atripennis Say 23-306   Ark. N.M.
bilineata Say 17-22   Conn.-Minn.
  lineata Melsh. 46-53 [21]
  mandibularis Melsh. 46-53 [21]
bridwelli Wellm. 12-38   Cal.
calceolata Guer. 29-136   N.A.
californica Wickh. 05-171   Cal.
cribraria Lec. 53-348   M.W.St.
cribricollis Lec. 53-348 Ind.-Tex.-Ariz.
  fuscipennis Lec. 53-349 [21]
  porosa Lec. 53-349 [21]
dichroa Lec. 53-346 Mont. Cr. Cal. ?Ar.
dubia Lec. 53-346   Cal. Ore.
dunniana Csy. 91-170   Tex.
flavida Lec. 53-349   N.M. Cal.
immaculata Say 17-22 Kan. Colo. N.M.
    [Ariz.
longicornis Horn 70-93   Ill. Fla.
lurida Lec. 53-345 Mex. Kan. Ar. Ore.
  decipiens Lec. 53-347 [21]
  rufa Duges 89-111 [21]
lutea Lec. 53-346   Kan. Ariz. Cal.
  pallens Lec. 53-346 [21]
martini Fall 07-257   N.Mex.
nemorensis Hentz 30-258   Fla.-Ind.
  bimaculata Melsh. 46-54 [21]
  ruficollis Beaur. 90-474 [21]
nigra Lec. 53-346   "Benicia"
nigripennis Lec. 53-347   N.M. Cal.
perforata Csy. 91-170   Tex.
piezata Fab. 94-104 Mex. Fla.-Ariz.-
  vittata Fab. 01-24 [21]   [Mont.
  texana Lec. 53-347 [21]
  v.bicolor Lec. 53-345   Kan. Colo.
    discolor Lec. 58-77 [21]   [Tex.
    stellaris Beaur. 90-465 [21]
  ?v.palliata Lec. 53-346 [21] L.Sup. Mont.
punctipennis Lec. 80-214   Ariz.
punctulata Lec. 53-347   Ga.
  v.flavipennis Uhler 55-418 [21] Va. Ind.
rufa Lec. 54-85   Tex. Mex.
  rubra Duges 70-166 [21]
schaefferi Blatch. 22-28   Fla.
scutellaris Lec. 53-347   Cal. Utah
sparsa Lec. 68-53   Colo. N.M.
sulcicollis Blatch. 10-1357   Ind.

[21] Revision of entire family by Denier—35. A few of our species are not mentioned, and the genus Pyrota is from Denier—34 where it was more completely treated. Other additions follow this revision.

vermiculata Schffr. 05-138   Utah
vigilans Fall 07-257   Cal.
vittigera Lec. 53-348 Ind.-Tex.-Ariz.
vittipennis Horn 75-155   Ariz.
zonitoides Duges 89-110 Mx. Guat. Tex.

Gnathium Kby. 18-425 [21]
francilloni Kby. 18-426 Mex. Tex. Ga.
  flavicolle Lec. 58-23 [21]
longicolle Lec. 58-77   Tex.
minimum Say 23-306 Mx. Kan.-Ark.-Ar.
  walckenæri Cast. 40-281 [21]
nitidum Horn 70-95 Mex. Cal. Ariz.
texanum Horn 70-94   Tex. Ariz.

SITARINI [21]

Tricrania Lec. 60-320 [21]
  (Horia Lec. 62-270) [21]
murrayi Lec. 60-320   Ore.
sanguinipennis Say 23-279 Conn.-Ind.

Tricraniodes Welim. 10-219 [21]
stansburyi Hald. 52-377   Utah

HORNIINI [21]

Hornia Riley 77-564 [21]
minutipennis Riley 77-564   Mo.

Leonidea Ckll. 00-11 [21]
  (Leonia Duges 89-211) [21]
anthophoræ Mickel 28-38   Colo.
neomexicana Ckll. 99-416 N.M. Kan.
  gigantea Welim. 11-16 [21]

HORIINI [21]

Cissites Latr. 04-154 [21]
  (Horia Fabr. 87-164, part) [21]
auriculata Champ. 92-372 C.A. Mx. Ar.
  maculata || Lec. & Horn 83-417 [21]

Meloinæ [21]

EPICAUTINI

Epicauta Redt. 45-621 [21]
  (Isopentra Muls. 58-106) [21]
  (Apterospasta Lec. 62-272) [12]
  (Macrobasis Lec. 62-272 [21]
abadona Skinner 04-217   Ariz.
alastor Skinner 04-217   Ariz.
albida Say 23-305 Mex. Kan. Colo.
  luteicornis Lec. 54-84 [21]
albolineata Duges 77-64 Mex. Ariz.
  duplicata Csy. 91-172 [21]
alphonsii Horn 74-38   Cal.
atrovittata Lec. 54-224 Tex.-Ariz.
batesi Horn 75-153   Ga.
borrei Duges 81-145 Mex. ?U.S.
  fumosa || Haag 80-40 [21]
  fumea Champ. 99-178 [21]

hoppingi Wellm. 12-35   Cal.
incommoda Horn 83-312   Cal.
   nunenmacheri Wellm. 12-36 [21]
infidelis Fall 01-303   Wash.-Cal.
insperata Horn 74-39   Cal.
lecontei Heyden 90-99   Mex. Tex.
   dichroa || Lec. 53-332 [21]
lugens Lec. 51-161   Cal.
maculicollis VanD. 29-130   Cal.
magister Horn 70-90   Cal. Ariz.
melæna Lec. 58-76   Ariz. Cal. L.Cal.
mœrens Lec. 51-216   Cal.
molesta Horn 85-111   Cal.
morosa Fall 01-301   Cal.
mutilata Horn 75-155   Mex. Ariz.
nigrocyanea VanD. 29-129   Colo.
nitidicollis Lec. 51-160   Cal. L.Cal.
nuttalli Say 23-300   Can. Kan. Colo.
   fulgifera Lec. 49-90 [21]   [Mont.
occipitalis Horn 83-312   Cal.
pilsbryi Skinner 06-217   Tex.
puberula Lec. 66-162   Ariz. Colo.
   variabilis Horn 85-107 [21]   Mex.
purpurascens Fall 01-302   Cal.
quadrimaculata Chevr. 34-79 C.A. Tex.
rathvoni Lec. 53-335   Cal.
refulgens Horn 70-91   Cal.
reticulata Say 23-305   Kan. Colo. N.M.
sphæricollis Say 23-299   Kan.-Wash.
   chalybea Lec. 51-160 [21]
   chalybeata Gemm. 70-124 [21]
stolida Fall 01-302   Cal.
tenebrosa Lec. 51-160   Cal.
ulkei Beaur. 89-212   Nev. Cal.
   lugubris Ulke 75-812 [21]
viridana Lec. 66-162   Kan.-Man.
vulnerata Lec. 51-159   Cal. Nev. Ore.
   angulicollis Duges 89-105 [21]   Mex.
  v. cooperi Lec. 54-18 [12] Cal.Wash.Ida.

Tetraonyx Latr. 05-204 [21]
   (*Jodema* Pasc. 62-57) [21]
albipilosa VanD. 29-127   Tex.
dubiosus Horn 94-440   L.Cal.
frontalis Chevr. 33-14   Mex. Tex.
  v.femoralis Duges 70-104 [21]
                [Mex. Ariz.
quadrimaculatus Fahr. 75-50   W.I.
   ruficollis Cliv. 95-14 [21]   [Fla.-S.C.

Pyrota Dej. 33-224 [21] [22]
akhurstiana Horn 91-42 Ariz.N.M.Mex.
bilineata Horn 85-115
          [Colo. Ariz. N.M. Tex.
concinna Csy. 91-174   Tex.
dakotana Wickh. 03-73   S.Dak.
discoidea Lec. 53-338   Tex.
dispar Germ. 24-171   S.A.
  v.brunneipennis Denier 34-60 [22]
   dispar Germ. 24-171 [22]   U.S.
divirgata Vill. & Pen. 67-15 Tex. Mex.
   nigrovittata Haag 80-51 [22]
   virgata Schffr. 05-177 [22]

engelmanni Lec. 48-91
        [Ind. Miss. Tex. N.M.
germari Hald. 43-303   Md.-Ga.
   mutata Gemm. 70-124 [22]
insulata Lec. 58-22   Tex. Mex.
invita Horn 85-114   Tex.
  v.limbalis Lec. 66-160 [22]   Va. Fla.
lineata Cliv. 95-14   Fla.-Tex.
mylabrina Chevr. 34-3
             [Ariz. N.M. Mex.
obliquefascia Schffr. 08-320   Ariz.
postica Lec. 66-160   Tex.-Ariz. Mex.
   maculata (Klug) Lac. 59-4 [22]
   plagiata Haag 80-49 [21]
   lacordairei Berg 81-303 [22]
punctata Csy. 91-173   Tex.-Ariz. Mex.
sinuata Cliv. 95-9   S.C. Fla. La.
   afzeliana Fabr. 01-78 [22]
tenuicostata Duges 77-60   Tex. Mex.
   dubitalis Horn 85-113 [22]
   vittigera || Lec. 58-22 [22]
   ? *rufipennis* Cr. 74-114 [22]
terminata Lec. 66-159   Mo.-Ut. L.Cal.
trochanterica Horn 94-439   L.Cal.

Pomphopœa Lec. 62-273 [21]
ænea Say 23-301   Pa. Ill. Tex.
   nigricornis Lec. 48-90 [21]
   filiformis Lec. 48-91 [21]
   tarsalis Bland. 64-71 [21]
polita Say 23-302   Fla.-S.C.
  v.femoralis Lec. 53-336 [21]   Fla. La.
   pedestris Harold 70-124 [21]
sayi Lec. 53-336   Can. Ct.-Ill. Tex.
   ænea var. Say. 24-228 [21]
   pyrivora Fitch 56-354 [21]
texana Lec. 66-161   Tex.
unguicularis Lec. 66-160   Ill. Ct.

Poreospasta Horn 67-139 [21]
polita Horn 67-139   Cal.
sublævis Horn 67-140   Cal.

Meloe Linn. 58-419 [21]
   TREIODONS Duges 70-102 [21]
barbarus Lec. 61-354   Is. off Cal.
lævis Leach 15-249   W.I. C.A. Mex.
   cordilleræ Chevr. 29-133   [N.M.
   sublœvis Lec. 54-84 [21]   [Ariz.
   tridentatus Jimenez 66-225 [21]
   tucci Pen. & Barr. 66-11 [21]
   opacus Mots. 73-48 [21]
   PROSCARABÆUS Schrank 81-
                   [225 [21]
   (*Cnestocera* Thoms. 59-124)[21]
afer Bland 64-70   Neb.
americanus Leach 15-251 Ga.Ind.Can.
  s.occidentalis VanD. 28-422 [21]
               [Id. Colo. Alb.
angusticollis Say 23-380 Can. Pa. Ind.
   rugipennis Lec. 53-328 [21]
californicus VanD. 28-426   Cal. Wash.
carbonaceus Lec. 66-155   Neb.
franciscanus VanD. 28-437   Cal. Nev.

---

[22] This genus is generally credited to Leconte (1862) but was amply validated by Dejean in 1833.

impressus Kby. 37-242   Can. Ind.
  americanus Brandt & Er. 32-118 [21]
  v.niger Kby. 37-241   Can.
mœrens Lec. 53-328   N.Y.
montanus Lec. 66-155   Mont.
opacus Lec. 61-354   Cal.
perplexus Lec. 53-329   Pa.
quadricollis VanD. 28-431   Cal.
strigulosus Mannh. 52-349   Alas. Cal.
tinctus Lec. 66-155   Neb.

Calospasta Lec. 62-273 [21]
  decolorata Horn 94-437   L.Cal.
  elegans Lec. 51-161   Cal.
    v.humeralis Horn 70-93 [1]   Cal.
    s.perpulchra Horn 70-92 [21]   Cal.
    s.cyanea VanD. 29-132 [21]   Cal.
  fulleri Horn 78-59   Cal.
  histrionica Horn 91-100   Cal.
  imperialis Wellman 12-37   Cal.
  mœsta Horn 78-59   Cal.
  morrisoni Horn 91-102   Cal.
  nemognathoides Horn 70-92 Cal. Ariz.
  schwarzi Wellman 09-23   Cal.
  sulcifrons Champ. 92-394   Mex. Ariz.
  viridis Horn 83-312   Colo. N.M.
  wenzeli Skinner 04-217   Ariz.

Pleurospasta Wellman 09-20 [21]
  mirabilis Horn 70-121   Ut. Ariz. Cal.

Phodaga Lec. 58-76 [21]
  alticeps Lec. 58-77   Mex. Ariz. Cal.

Negalius Csy. 91-175 [21]
  marmoratus Csy. 91-175   Tex. Ariz.
                           [L.Cal.

Tegrodera Lec. 51-159 [21]
  erosa Lec. 51-159   Cal. L.Cal.
    v.extincta Beaur. 90-493 [21]
    v.aloga Skinner 03-168 [21]   Ariz.
    v.inornata Blais. 18-334 [21]   Ariz.
  latecincta Horn 91-44   Cal.

Eupompha Lec. 58-21 [21]
  fissiceps Lec. 58-21   Tex.

Brachyspasta VanD. 28-451 [21]
  wickhami VanD. 28-452   Colo.

Cordylospasta Horn 75-152 [21]
  fulleri Horn 75-152   Nev.

Gynæcomeloe Wellman 10-217 [21]
  opacus Horn 68-139   Cal.
  parvicollis VanD. 28-450   Cal.

Cysteodemus Lec. 51-158 [21]
  armatus Lec. 51-158   Cal. Ariz.
  wizlisenoi Lec. 51-158   Mex. N.M. Ariz.

Megetra Lec. 59-127 [21]
  cancellata Brandt & Er. 32-141   Mex.
    hoegei Duges 89-39 [21]   [U.S.
  vittata Lec. 53-330   Mex. N.M. Ariz.
           •   •   •

[21] Denier—35.

Epicauta
  elongatocalcarata Mayd. 34-328   Ida.
  piceiventris Mayd. 34-327   Utah
  rehni Mayd. 34-329   Ariz.
  excavatifrons Mayd. 34-330   Fla.
  mutchleri Mayd. 34-331   Ariz.
  diversipubescens Mayd. 34-333   N.M.
  crassitarsis Mayd. 35-72   Ariz.

Macrobasis
  hirsutipubescens Mayd. 34-334   Ariz.
  maculifera Mayd. 34-335   Ariz.

PYTHIDÆ

Cariderus Muls. 59-46 [23]
  (*Rhinosimus* Latr. 02-192) [23]

Mycterus Clairv. 98-125 [24]
  canescens Horn 79-336   Cal. Ore.
  elongata Hopping 35-77   Cal.
  concolor Lec. 53-235
            [Cal.-B.C. Colo. N.M.
  scaber Hald. 43-303   Mass.-Va.
  quadricollis Horn 74-142   Cal. N.M.

PYROCHROIDÆ

Ischalia Pasc. 60-54
  EUPLEURIDA Lec. 73-335 [25]
  californica VanD. 38-192   Cal.

ANTHICIDÆ

Notoxus
  visaliensis Blais. 36-144   Cal.
  sparsus Lec. 59-284 [26]   Cal. Ariz.
  conformis Lec. 51-152 [26]   Cal.
  obesulus Blais. 36-146   B.C.

CEBRIONIDÆ

Cebrio
  convexifrons Knull 35-189   Ckla.

[23] Hopping—35.
[24] Transferred from Melandryidæ. Hopping—35.
[25] Van Dyke—38.
[26] Valid species. Blaisdell—36.

PLASTOCERIDÆ [27]

Plastocerus
maclayi Sloop 35-17   Cal.
pullus Sloop 35-18   Cal.

Euthysanius
brevis Sloop 35-19   Cal.

ELATERIDÆ

Adelocera
mexicana Cand. 57-70 [1]   Fla. Mex.
nobilis Fall 32-58 [1]   Ariz.
  mexicana of Lec.[1]

Conoderus Esch. 29-31
  (*Monocrepidius* Esch. 29-31)[3]
  HETERODERES Latr. 34-155 [1]
suturalis Lec. 53-482 [1]   Ala. Ind.
fuscosus Blatch. 25-163   Fla. Ga.
  fucosus of VanDyke [1]
  planidiscus Fall 29-56[1]

• • •

exsul Shp. 77-470 [3] N.Zeal. Hawaii, Cal.
rudis Brown 33-174   Ala.
browni Knull 38-19   Tex.
perversus Brown 33-173   Fla.

Aeolus
melillus Say [4]   Ind. Ariz.
  elegans of authors [4]
  dorsalis || Say 23-167 [4]
  s. comis Lec. 53-484 [4]   B.C. Alb.
  s. marginicollis Horn 71-308 [4]
                    [Sask. Man. Ore.
  (*elegans* Fabr., *not North American*) [4]
  (*circumscriptus* Germ., *not N. Amer.*)[4]

Limonius
canus Lec. 53-433 (Pheletus) [1]   Cal.
discoideus Lec. 61-348 [1]
                    [Wash. Cal. Rky.Mts.
ovatus Knull 34-9   Cal.
griseus Beauv. 05-214 [4]   Ct.-Ind. Ont.
  interstitialis Melsh. 46-215 [4]
rudis Brown 33-175   Ont. Ind.
rectangularis Fall 34-30   Tex.
flavomarginatus Knull 38-20   Ohio
pilosulus Cand. 91-149 [5]   Cal.
  pilosus || Lec. 53-432 [1]
æger Lec. 53-431   N.B.-Alb.-Pa.
  knulli Fall 33-229 [1]

basilaris Say 23-172   E.U.S. Fla. Mo.
  v.semiæneus Lec. 53-432 [1] Ga. Fla.
insperatus Brown 33-175   Cal.
sinuifrons Fall 07-227   N.Mex.
  ovatus Knull 34-9 [1]
ectypus Say 39-167 [5]   Pa. Me.
agonus Say 39-171 [5]   Pa.

Elathous
brevicornis Fall 34-30   Cal.
brunnellus Fall 34-31   Cal.

Athous
rufiventris Esch. 22-71
  ferruginosus Esch. 29-33 [6]
paradisus Knull 34-10   Ariz.

Ludius
ochreipennis Lec. 63-85 [7]
                    [B.C. Y.T. N.W.T.
watsoni Brown 36-179 [7]   Que.
hoppingi VanD. 33-434 [7]
                    [B.C. Wash. Y.T.
bipunctatus Brown 36-180 [7]   B.C. Alb.
exclamationis Fall 10-135 [7]   Cal.
hieroglyphicus Say 39-172 [7]   N.H.-Pa.-
  bicinctus Cand. [7]   [Que.-Man.
  ctenicerus G. & H. 69-1577 [7]
pudicus Brown 36-183 [7]   B.C. Alb.
propola Lec. 53-437 [7]
  s.propola (s.str.) [7]   N.S.-L.Sup.
  furcifer Lec. 53-438 [7]
  s.columbianus Brown 36-185 [7]   B.C.
pallidipes Brown 36-185 [7]   Cal.
californicus Brown 36-186 [7]   Cal.
  nubilis || Lec. 53-438 [7]
candidus Brown 36-186 [7]   Cal.

• • •

cribrosus Lec. 53-443 [5] [8]   Cal.
colossus Lec. 61-348 [5] [8]   Cal.
maurus Lec. 53-444[5] [8]   Cal.-B.C.
æthiops Hbst. 06-70 [8]   Md. Tenn.
  nigrans Cast. 40-241 [8]
  depresssus G. & H. 69-1574 [8]
?signaticollis Melsh. 46-216 [8]
?subcanaliculatus Mots. 59-375 [8]
?rufipes Mots. 59-377 [8]

• • •

spinosus Lec. 53-447 [9]   Newf.-Ont.
stricklandi Brown 35-219 [9]   Ont.-Alb.
crestonensis Brown 35-219 [9]   B.C. Ida.
umbricola Esch. 29-34 [9]   Alas. B.C.
  rudis Mots. 59-376 [9]
varius Brown 35-220 [9]   Wash.
volitans Esch. 29-34 [9]   Alas. B.C.
vulneratus Lec. 63-86 [9]   N.S.-N.C.

• • •

[27] Reduced to subfamily of Elateridæ by Sloop
—35.
[1] Fall—34.
[2] Used as valid by Fall—34.
[3] Graves—38.
[4] Brown—33.
[5] As distinct species, Fall—34.

[6] Van Dyke in litt.
[7] Revision of propola-group, Brown—36.
[8] Revision of cribrosus-group, Brown—35.
[9] Revision of the volitans-group, Brown—35.

triundatus Rand. 38-12 [10]
       [N.B.-N.H.-B.C.
nebraskensis Bland 63-355 [10] B.C. Mont.
tigrinus Fall 01-306 [10]   Cal.
  trigrinus Brown 36-107

• • •

medianus Germ. 43-71 [11]   N.S.-L.Sup.-
  rubidipennis Lec. 53-437 [11] [Mass.
bombycinus Germ. 43-70 [11]
       [Ore. B.C. Alb.
fallax Say 39-605 [11]   Que. L.Sup.
viduus Brown 36-103 [11]   B.C.
semiluteus Lec. 53-445 [11]   Cal.
mirabilis Fall 01-306 [11]   Cal.
?elegans Cand. 82-97 [11] [12]   Cal.
  candezei Leng 18-205 [11] [12]
  (*sericeus* Gelb., *not North American*)[13]
  (*tesselatus* Linn., *not No. American*)[11]

• • •

splendens Zieg. 44-44 [14] Que. Ont. Mass.
  metallicus G. & H. 69-1580 [14] [Pa.
æripennis Kby. 37-150 [12] [14]
  s.æripennis (s.str.) [14]   Alas.-Wash.-
  tinctus Lec. 59-85 [12] [14]   [Wyo.-
  elegans of Schw. 07-225 [10] [14] [NWT
  s.destructor Brown 35-129 [14] [15]
       [Alas.-N.D.
montanus Brown 35-129 [14]   B.C.-Ore.
appropinquans Rand. 38-5 [14]
       [N.S.-Me.-Wis.
darlingtoni Brown 35-131 [14]   N.H. Me.
semimetallicus Walk. 66-325 [14]
       [B.C.-Ida.-Alb.
carbo Lec. 53-439 [14]   Ore.-Colo.-Mont.
lateralis Lec. 53-439 [14]   B.C. Ore.
imitans Brown 35-134 [14]   Cal.
pruininus Horn 71-320 [14] Cal.-B.C.-Neb.
  pruinosus Schw. 07-226 [14]
  pruinosulus Schw. 07-316 [14]
  noxius Hyslop 14-69 [14]
?confluens Gehl. 30-80 [14]   Asia, Alas.
?quadrivittatus Walk. 66-325 [14]   B.C.

• • •

nitidulus Lec. 53-439 [16]   N.S.-Alb.
rufopleuralis Fall 34-188 [16]
       [N.B.-Mass.-Man.
aratus Lec. 53-438 [16]   N.S.-L.Sup.
(*nigricornis* Panz., *not No. Amer.*)[16]

• • •

inflatus Say 25-392 [17]   Que.-S.C.-Ind.
  metallicus Say 25-392 [17]
glaucus Germ. 43-76 [17]   B.C.-Cal.-Ut.
  similissimus Mots. 59-374 [17]

• • •

callidus Brown 36-135 [17]   B.C. Ida.
inutilis Brown 36-136 [17]   Cal.
cruciatus Linn. 58-404 [18]   Eur. N.A.
  s.pulcher Lec. 53-440 [18]   N.S.-Ont.
  s.festivus Lec. 57-46 [18] B.C.-Ore.-Man.

• • •

edwardsi Horn 71-324 [19]
  s.edwardsi (s.str.) [19]   Cal. Nev.
  ater || VanD. 32-430 [19]
suckleyi Lec. 57-46 [19]
  s.suckleyi (s.str.) [19]   B.C. Ore.
  s.olympiæ VanD. 32-431 [19]   Wash.
  s.morulus Lec. 63-85 [19]   Alas.-Ore.
       [Y.T.-Alb.
  brunnipes Bland 64-67 [19]
sexualis Brown 35-8 [19] Sask. Alb. Wyo.

• • •

semivittatus Say 23-113 [20]   Mo.-Dak.-
       [Colo.
blanditus Brown 36-13 [20]   Cal.
sexguttatus Brown 36-14 [20]   Ore.
trivittatus Lec. 53-443 [20]   Ga.-N.C.
deceptor Brown 36-16 [20]   Cal.
funereus Brown 36-16 [20]   B.C.
castanicolor Fall 34-34 [23]   N.Mex.
?oblongoguttatus Mots. 59-373 [20]   Cal.
?tristis Cand. 63-172 [20]   Vanc.Id.
?fusculus Lec. 63-48 [20]   Cal.
  angustulus || Mots. 59-373 [20]

• • •

    CORYMBITES Latr. 34-150 [21]
kaweana Fall 37-31   Cal.
  rufipennis || Fall 10-134 [21]
conjungens Lec. 53-440 [1]   Cal.
  præses Cand. 65-28 (Drasterius)[1]
elegans Cand. 82-97 [2*]
  candezei Leng 18-205 [22]
æripennis Kby. 37-150 [22]
  appropinquans Rand. 38-5 [22]
  tinctus Lec. 59-85 [22]
  semimetallicus Walk. 66-325 [22]
  elegans Leng 20-170 (in error) [22]
rufopleuralis Fall 34-188   Que.-Mass.
       [Mich.
nigricans Fall 10-135 [23]   Cal.
rotundicollis Say 25-259 [23] Vt. Pa. Ind.
  diversicolor Esch. 29-34 [1]   [Cal.
castanicolor Fall 34-34   N.Mex.
Eanus Lec. 62-171 [1]
  (*Paronomus* || Kies. 63-303) [1]
striatipennis Brown 36-248   B.C. Ore.
maculipennis Lec. 66-85 [23]   Lab. Can.
  pictus Cand. 63-177 [1]
hatchi Lane 38-188   Wash.
estriatus Lec. 53-434
  subarcticus Brown 30-163 [1]   Que.
decoratus Mannh. 53-434   Can.
  parvicollis Mannh. 53-229 [15] Lapland

[10] Revision of the triundatus-group, Brown—36.
[11] Revision of the fallax-group, Brown—36.
[12] See end of genus for corrected synonymy.
[13] Schenkling—27.
[14] Revision of the æripennis-group, Brown—35.
[15] Brown—36.
[16] Revision of the nitidulus-group, Brown—36.
[17] Revision of the inflatus-group, Brown—36.

[1] Fall—34.
[18] Revision of the cruciatus-group, Brown—35.
[19] Revision of the edwardsi-group, Brown—35.
[20] Revision of the semivittatus-group, Brown—36.
[21] Fall—37.
[22] By compiler; see also Brown—33 and 35.
[23] Valid species, Fall—34.

Hypnoidus
CRYPTOHYPNUS Esch. 36-105
valens Fall 34-18   Cal.

* *

manki Fall 34-19   Mont.

Dalopius Esch. 29-34 [24]
* (Dolopius of authors) [24]

virginicus Brown 34-38	W.Va.
cognatus Brown 34-38	N.S.-Ont.-W.Va.
vagus Brown 34-66	N.B.-Man.-W.Va.
insolitus Brown 34-67	Ont. Que.
fuscipes Brown 34-68	Que.
pennsylvanicus Brown 34-68	Pa.
vernus Brown 34-69	Que.-Man.
brevicornis Brown 34-69	Que. Ont.
agnellus Brown 34-70	Que. N.B.
gentilis Brown 34-71	Ont. Que.
parvulus Brown 34-71	Sask. Man.
pallidus Brown 34-87	N.B.-Alb.
gartrelli Brown 34-89	B.C.
luteolus Brown 34-89	Cal.
corvinus Brown 34-90	B.C.
usitatus Brown 34-90	Cal.
insolens Brown 34-91	B.C.
asellus Brown 34-91	B.C. Alb.
jucundus Brown 34-92	Cal.
gracilis Brown 34-92	Cal.
tularensis Brown 34-93	Cal.
dentatus Brown 34-94	Cal.
inordinatus Brown 34-94	Alb.
mirabilis Brown 34-95	Que.-Alb.
plutonicus Brown 34-95	Cal.
incomptus Brown 34-96	Cal.
lutulentus Brown 34-96	Cal.
validus Brown 34-102	Cal.
spretus Brown 34-103	B.C.
vetulus Brown 34-103	Cal.
partitus Brown 34-104	Cal.
effetus Brown 34-104	Cal.
suspectus Brown 34-104	B.C.
fucatus Brown 34-105	Alb. B.C.
tristis Brown 34-105	B.C.
manipularis Brown 34-107	Cal.
ignobilis Brown 34-107	Cal.
improvidus Brown 34-108	Cal.
insulanus Brown 34-108	B.C.
invidiosus Brown 34-109	Cal.
maritimus Brown 34-109	B.C.
mutabilis Brown 34-110	Cal.
?californicus Mann. 43-243 [24]	Cal.

lateralis ‖ Esch. 29-34 [24]
pauperatus Cand. 63-434 (Sericoso-
       [mus [24]
?sellatus Mannh. 52-328 [24]   Alaska
?pauper Lec. 53-458 [24]   Atl. N.A.
?subustus Lec. 53-458 [24]   Cal.
?macer Lec. 57-47 (Agriotes) [24]   Ore.
?simplex Mots. 59-378 [24]   Cal.
?sericatus Mots. 59-379 [24]   Cal.
?nevadensis Lec. 84-17 (Agriotes) [24]

[24] Revision of genus, Brown—34.

Agriotes
tardus Brown 33-177	Alb. B.C.
quebecensis Brown 33-177	N.C.-Ont. Me.
opaculus Lec. 59-85	Cal. Id. Wyo.
montanus Lec. 84-18 [15]	
arcanus Brown 33-178 [25] [26]	
isabellinus Melsh. 46-218 [25]	Pa.
oblongicollis Melsh. 46-218 [25]	M.St. Ct.
	[Ind.

Agriotella Brown 33-179
bigeminata Rand. 38-37 [27]	N.S.-Ont.-
	[Mass.
occidentalis Brown 33-180	B.C. Alb.
californica Schffr. 17-42 [27]	Cal.
columbiana Brown 33-182	B.C. Cal.

Drasterius [28]
dorsalis Say 23-167 [28]	N.A.
comis Lec. 53-484 [28]	N.A.
livens Lec. 53-484 [28]	Cal. Mex.
amabilis Lec. 53-485 [28]	M.St. S.St.
fretus Csy. 85-171 [28]	
nigriventris Schffr. 17-41 [28]	Tex.
scutellatus Schffr. 17-41 [28]	Tex.
debilis Lec. 78-405	Fla.
incongruus Fall 34-12 [29]	Que.

Ampedus Germ. 44-153
(Elater of authors) [30]
obscurus Knull 38-97	Nev.
varipilis VanD. 32-306	Cal.
s.columbianus Brown 33-136	B.C.
	[VanId.
quebecensis Brown 33-137	Que.
evansi Brown 33-137	N.B.-Ont.
areolatus Say 23-167	Ont.
pusio Germ. (not Lec. or Cand.) [31]	
luteolus Lec. 53-471	Ont. N.C.
pusio Lec. and Cand. (Elater) [31]	
melanotoides Brown 33-134	Ont. Pa.
laurentinus Brown 33-135	Que. Ont.

Blauta
falli Brown 36-251   Fla.

Megapenthes
solitarius Fall 34-15 [32]	N.Y.
granulosus Melsh. 44-159 [33]	Fla.-Conn.
(sturmii Germ., not North American) [29]	
dolosus Brown 33-140	Cal.
stigmosus Lec. 53-472 [31]	N.B.-B.C.
caprella Lec. 57-47 [31]	Ore. B.C.
s.californicus Brown 33-140	Cal.

[25] Valid species, Brown—36.
[26] As synonym of isabellinus, Fall—34.
[27] Transferred to new genus from Betarmon, Brown—33.
[28] These species belong here with some others, Fall—34.
[29] Fall—34.
[30] Implied by Knull—38 and Brown—33.
[31] Brown—33.
[32] Neither this species, lepidus Lec. tarsalis Schffr., nor illinoiensis VanD. belong in this genus, Fall —34.
[33] Valid species, Fall—34.

Anchastus
 fumicollis Fall 34-17   Fla.
 longulus Lec. 78-404 [33]   Fla.
 uniquus Knull 38-98   Tex.
 subdepressus Fall 34-18   Ariz.

Melanotus [34]
 longulus Lec. 53-473 [34]   Cal.
 oregonensis Lec. 53-480 [34] Cal.-B.C.-Ut.
 franciscanus VanD. 32-332 [34]   Cal.
 variolatus Lec. 61-377 [34]   Cal.

Horistonotus
 pallidus Fall 34-21   Cal.
 fidelis Fall 34-21   Cal. Nev.
 s.fuscus Fall 34-22   Ariz.

Esthesopus
 flavidus Fall 01-240 (Horistonotus)[29]
                        [Cal.
 indistinctus Fall 34-23   Cal.

MELASIDÆ

Melasis
 rufipalpis Chevr. 35-193 [35]   Guat. Mex.
                            [Ariz.

BUPRESTIDÆ

Chrysophana
 placida
  a.cœrulans Obenb. 24-100 [1] [2]
  v.strandi Obenb. 36-106   Cal.
  v.cupriola VanD. 37-105   Cal.
  s.conicola VanD. 37-105   Cal.

Polycesta Sol. [3] (for 32-281 read 33-281)[4]
 * 9306, for 78-144 read 85-144 [4]
 angulosa Duval 57-62 [3]   Fla. Ala.
   obtusa Lec. 58-68 [3]
 velasco Gory 38-5 [3]   Tex.-Cal. L.Cal.
                            [Mex.
 arizonica Schffr. 06-21 [3]   Cal.-Ariz.
 elata Lec. 58-68 [3]   Tex.
 ?cavata Lec. 58-68 [3]   ?Ala.
 cyaneous Chamb. 33-41 [3]   Cal.
 tularensis Chamb. 38-445   Cal.
 californica Lec. 57-45 [3]   Cal. Ore.
   cribrana Mots. 59-182 [3]
   bernardensis Obenb. 24-35 [3]

Acmæodera
 resplendens VanD. 37-106   Ariz.
 delumbis Horn 94-378 [5]
 robusta
  s.dubolsi Cazier 38-138   Cal.

uvaldensis Knull 36-73   Tex.
yumæ Knull 37-301   Ariz.
adenostomæ Cazier 38-137   Cal.
nigrovittata VanD. 34-61 *   Cal.
simulata VanD. 37-108   Cal.
mariposa
 s.bernardino VanD. 37-109   Cal.
papagonis Duncan 34-231   Ariz.
vanduzeei VanD. 34-64   L.Cal.
varipilis VanD. 34-62   G. of Cal. Tex.
                            [Ariz.
vulturei Knull 38-228   Cal.
constricticollis Knull 37-301   Ariz.
biedermanni Skinner 03-239 [7]   Ariz.
lineipicta Fall 31-81 [7] [8]   Ariz.
sabinæ Knull 37-15   Ariz.
rossi Cazier 37-115   Cal.
humeralis Cazier 38-12   Cal.
jaguarina Knull 38-135   Ariz.
mimicata Knull 38-136   Ariz.

Tyndaris
 * balli Knull 37-302   Ariz.

Paratyndaris Fisher 19-92 *
 *   (*Tyndaris* of authors)[10]
 anomalis Knull 37-252 *   Nev.
 tucsoni Knull 38-21   Ariz.
 albofasciata Knull 37-253 *   Ariz.
 barberi Skinn. 03-238 *   Nev. Ariz.
 cincta Horn 85-147 *   Tex.
 acaciæ Knull 37-254 *   Tex. N.M.
 olneyæ Skinn. 03-236 *   Nev. Ariz.
 prosopis Skinn. 03-237 *   Ariz.
 coursetiæ Fisher 19-93 *   Ariz.
 quadrinotata Knull 38-22   Ariz.
 chamæleonis Skinn. 03-237 *   Tex.
 suturalis Fall 34-193 *   Fla.

Thrincopyge
 alacris
  v.strandi Obenb. 36-104   Tex.

Agæocera
 gigas Cast. & Gory 39-2 [11]   Mex. Ariz.

Hippomelas
 pacifica Chamb. 38-446   Cal.
 9313, for 76-288 read 83-288 [4]

Texania
 strandi Obenb. 36-109   U.S.A.[12]

Psiloptera
 riograndei Knull 37-16   Tex.

---

[29] Fall—34.
[33] Valid species, Fall—34.
[34] Revision of California species, Fall—34.
[35] Fall—33.
[1] Obenberger—36.
[2] Van Dyke—37.
[3] Revision of genus, Chamberlin—33.
[4] Chamberlin in litt.
[5] Valid species or variety, Duncan—34.

[6] Possibly a synonym of lineipicta, Fall—34.
[7] Cited as distinct species, Fall in litt.
[8] Listed as a synonym of biedermanni, Van Dyke —34.
[9] Revision of genus, Knull—37.
[10] Obenberger—Col. Cat. p. 84.
[11] Fall—34.
[12] "U.S.A.: Fort Madison."—Obenberger—36.

Dicerca
9333, for 23-13 read 23-163 [4]
hesperoborealis Hatch & Beer 38-151
[Wash.
tenebrica Kby. 37-155
prolongata Lec. 59-194 [8]
sexualis Cr. 73-87 Cal.-Wash.
californica Cr. 73-87 [8] [13]
obscura Fab. 81-274
mutica Lec. 59-196 [8]

Hesperorhipis
(*Hesperohipis* of 2nd suppl. p. 29) [14]
mirabilis Knull 37-303 Ariz.
hyperbolus Knull 38-137 Ariz.

Buprestis
intricata Csy. 09-118 [15] Cal.
maculativentris Say. 24-272
v.subornata Lec. 59-208 [16]
rusticorum Kby. 37-151 [15] B.C.-Cal.-Ar.
nuttalli Kby. 37-152
alternans Lec. 59-207 [16]
consularis Gory 40-120 [16]
langi Mannh. 43-237 [15] Alas.-Ore.
murrayanæ R.Hopp. 34-174
contortæ || R.Hopp. 33-84 [16] [17]

Xenorhipis
osborni Knull 36-73 Tex.

Melanophila Esch. 29-9 [18]
* MELANOPHILA (s.str.) [18]
consputa Lec. 57-44 [16] Cal. Ariz.
monochroa Obenb. 28-209 [18]
isolata Obenb. 28-209 [16]
notata Cast. 37-4 [15] Fla.-Mex.-M.St.
luteosignata Mannh. 37-70 [18] [Cuba
s.elegans Sloop 37-7 Ariz.
opaca Lec. 59-213 [18] Fla. Ga.
notata Cr. 73-89 [16]
acuminata DeG. 74-133 [18] No.Holarctic
acuta Gmel. 88-1939 [18]
morio Fahr. 92-210 [16]
appendiculata Fahr. 92-210 [18]
pecchiolii Cast. & Gory. 37-33 [18]
longipes Say 23-164 [16]
immaculata Mannh. 37-70 [18]
occidentalis Obenb. 28-209 [18] Cal.
atropurpurea Say 36-213 [16] Ariz.-Ut.-
[Ark.
PHÆNOPS Lac. 57-47 [18]
obtusa Horn 82-106 [18] Ga.
intrusa Horn 82-105 [16] Pac.Cst.
æneola Melsh. 46-146 [18] E.&M.&S.Sts.
metallica Melsh. 46-146 [18]
carolina Manee 13-164 [18] N.C.

[13] This synonymy conflicts with that given by Obenberger—Col. Cat. p. 111.
[14] Fall in litt.
[15] Distinct species, R.Hopping—33.
[16] R.Hopping—33.
[17] Casey's name was spelled contorta and might therefore be considered not to invalidate contortæ of Hopping.
[18] Revision of genus, Sloop—37.

drummondi Kby. 37-159 [16] Cal.-Wash.
v.abies Champl. & Kn. 23-105 N.B. [18]
abietis Obenb. 30-441 [18]
gentilis Lec. 63-42 [16] Pac.Cst.
prasina Lec. 60-254 [18]
pini-edulis Burke 08-117 [18] Colo. Ut.
[Ariz.
fulvoguttata Harris 29-2 [16] N.E.N.A.
octospilota Cast. & Gory 37-4 [18]
croceosignata Cast. & Gory 37-5 [18]
decolorata Cast. & Gory 37-5 [16]
guttulata Hamilt. 89-138 [18]
cauta Dej. 33-89 [18]
drummondi Kby. 37-159 [16] Ore.-Ariz.
umbellatarrum Kby. 37-159 [18]
guttulata Mannh. 53-221 [16]
a.tristicula Obenb. 28-209 [18]
?lecontei Obenb. 28-210 [18] Cal.
arcuata Fall 31-83 [16] Ariz.

XENOMELANOPHILA Sloop
[37-18
miranda Lec. 54-83 [16] N.Mex.

Anthaxia
viridifrons
californica VanD. 18-54 [16] Cal.
a.basicyanea Obenb. 36-134 N.J. Pa.
a.embrikaria Obenb. 36-134 Can.-
[N.J.-Ind.
vandykeana Obenb. 36-135 Cal.
embrik-strandella Obenb. 36-135 Cal.
serripennis Obenb. 36-136 Iowa
quercata
a.floridana Obenb. 36-134 Fla.
cyanella
a.rossi Obenb. 36-134 Ky. Pa.

Chrysobothris
atrifasciata Lec. 73-332
atrofasciata of Leng Cat. p. 182 [19]
platti Cazier 38-14 Cal.
alleni Cazier 38-15 Ariz.
schæfferi Obenb. 34-649 L.Cal.
thoracica || Schffr. 05-128 [20]
scotti Chamb. 38-11 N.Mex.
chiricahuæ Knull 37-37 Ariz.
kelloggi Knull 37-36 N.Mex.
iris VanD. 37-110 Utah
grandis Chamb. 38-14 Ore.
parapiuta Knull 38-138 Ariz.
arizonica Chamb. 38-13 Ariz.
oregona Chamb. 34-38 Ore.
grindeliæ VanD. 37-111 Cal.
canadensis Chamb. 34-37 Alb. Ore.
cupreohumeralis VanD. 34-65 Tex.
boharti VanD. 34-89 Cal.
acaciæ Knull 36-105 Tex.
9472, for 86-108 read 86-101 [19]
bisinuata Chamb. 38-13 Cal.
calcarata Chamb. 38-12 Ariz.
planomarginata Chamb. 38-10 Ore.
Knowltonia Fisher 35-117 [21]
* biramosa Fisher 35-118 Utah

[19] Chamberlin in litt.
[20] Obenberger—34.
[21] Considered a synonym of Chrysobothris by Cazier—38.

Engyaulus Waterh. 89-50
* pulchellus Bland 65-382   Colo. Ariz.
  pinalicus Wickh. 03-69 [22]   [N.M.

Agrilus
  santaritæ Knull 37-39   Ariz.
  arizonicus Obenb. 36-139   Ariz.
    strigicollis ‚' Fall 12-41 [1]
  sapindi Knull 38-139   Tex.
  shoemakeri Knull 38-99   Ariz.
  parkeri Knull 35-189   Ariz.
  neabditus Knull 35-190   Ariz.
  exhauchucæ Knull 37-305   Ariz.
  apachei Knull 38-139   Ariz.
  9542, for 25-241 read 25-251 [19]
  cercidii Knull 37-304   Ariz.
  esperanzæ Knull 35-96   Tex.
  arizonus Knull 34-11   Ariz.
  osburni Knull 37-38   Ohio
  viridescens Knull 35-97   Tex.
  parapubescens Knull 34-68   Ariz.
  wenzeli Knull 34-333   Ariz.
  rubicola Abeille 97-4 [23]   Eur. E.N.A.
    viridis of authors [23]
    proximus Rey 91-19 [23]
    obtusus ‚' Abeille 97-5 [23]
    antiquus Gavoy 26-15 [23]
    chrysoderes || Bedel 21-204 [23]
    fagi || Glover 78-A33 [23]
    politus || Weiss 14-438 [23]
  v.communis Obenb. 24-41 [23]   Eur.
                                    [E.N.A.
  neoprosopidus Knull 38-99   Tex.
  cavifrons Waterh. 89-189 [24] Mex. Ariz.

Paragrilus Saund. 71-127 [25]
  (*Clinocera* |¡ Deyr. 64-116)[25]
  (*Rhœboscelis* Lec. 63-82, part)[25]
  lesueuri Waterh. 89-126 [25]   S.A. C.A.
                                    [?Ariz.
  tenuis Lec. 63-82 [25]   N.Y.-Fla.-Ill.
  texanus Schffr. 04-211 [25]   Tex.

Taphrocerus Sol. 33-314 [26]
* howardi Obenb. 34-42   Fla.
  gracilis Say 25-253 [26]   E.U.S.-Ia.-Ariz.
    alboguttatus Mannh. 37-120 [26]
    cylindricollis Kerr 96-312 [26]
    texanus Kerr. 96-312 [26]
  nicolayi Obenb. 24-55 [26]   N.Y. Mass.
  albonotatus Blatch. 19-29 [26]   Fla.-N.C.
  floridanus Obenb. 34-48   Fla.
  lævicollis Lec. 78-403 [26]   Fla. Ala.
  agriloides Cr. 73-96 [26]   Fla.-Tex.
  puncticollis Schw. 78-363 [26]   Fla. Ala.

Mastogenius
  subcyaneus
    s.crenulatus Knull 34-334 N.J.-Fla.-
                                     [La.

[1] Obenberger—36.
[19] Chamberlin in litt.
[23] Obenberger—35.
[23] Obenberger—Col. Cat. p. 152.
[24] Knull—37.
[25] Revision of the genus, Obenberger—35.
[26] Revision of genus, Obenberger—34.

DRYOPOIDEA [1]

LIMNICHIDÆ [1]

Limnichinæ [1]

Limnichites [1]

Lutrochus [1]

Cephalobyrrhinæ [1]

Throscinus [1]

DRYOPIDÆ [1]

Dryopinæ (*Pelonominæ*) [1]

Pelonomus [1]

Oberonus [1]

Dryops [1]

Helichus Er. 47-510 [1][2]
  (*Dryops* of Leach 17-88) [2]
  (*Parygrus* Er. 47-510)[2]
  (*Pachyparnus* Fairm. 88-338)[2]
  productus Lec. 52-43 [2]   Cal.
  immsi Hntn. 37-318   Cal. Ariz. Tex.
  propinquus Hntn. 35-68   L.Cal.
  confluentus Hntn. 35-71 [2]   Ga.-Ariz.
    longulus Musg. MS [2]
  lithophilus Germ. 24-88 [2]   Can.-Fla.-Ia.
  basalis Lec. 52-43 [2]   Mass.-Ga.
    falli Musg. MS [2]
  fastigiatus Say 24-275 [2] Mass.-Fla.-Kan.
  striatus Lec. 52-43 [2]   Que.-Cal.-B.C.
    v.foveatus Lec. 52-43 [2]
    columbianus Brown 31-118 [2]
  triangularis Musg. 35-143 [2]   Ariz. Tex.
  suturalis Lec. 52-43 [3]   Cal. C.A.
    gilensis Lec. 52-43 [3][4]
    æqualis Lec. 54-81 [3][4]
                                • • •
    elmoides Shp. 82-121 [4]
                                • • •
  puncticollis Shp. 82-121 [5]   Ariz. ?C.A.

PSEPHENIDÆ [1]

Psepheninæ [1]

Psephenus [1]

[1] Notes on relationships of families, subfamilies,
  and genera, Hinton—39.
[2] Revision of genus, Musgrave—35.
[3] Hinton—36.
[4] Hinton—37.
[5] Hinton—84.

Eubrianacinæ [1]

Eubrianax [1]

ELMIDÆ *(Helmidæ)* [1]

Larinæ [1]

Lara [1]

Phanocerus [1]

Elminæ [1]

ELMINI [1]

Elsianus [1]

Stenelmis Duf. 35-158 c
  nubifera Fall 01-238   Cal. Ore. Wash.
  sexlineata Sandn. 38-663   Kan. Tex.
  crenata Say 24-275   N.B.-Ala.-Tex.
    sordida Mots. 59-51 c
  exigua Sandn. 38-669   Ark.
  beameri Sandn. 38-671   Mo.-Ark.
  lateralis Sandn. 38-672   Pa.-Va.-Ark.
  concinna Sandn. 38-674   Que.-N.C.
  tarsalis Sandn. 38-675   Ont.-W.Va.-Okla.
  knobeli Sandn. 38-677   Ark.
  bicarinata Lec. 52-44   Vt.-Tex.
  exilis Sandn. 38-680   Ark.
  mera Sandn. 38-682   Que.-N.C.-Ark.
  douglasensis Sandn. 38-685   Mich.
  grossa Sandn. 38-686   Miss. La. Ark.
  parva Sandn. 38-688   Okla. Tex.
  fuscata Blatch. 25-164   Fla.
  hungerfordi Sandn. 38-690   Fla.
  humerosa Mots. 59-50   Mass.-S.C. Tenn.
  linearis Zimm. 69-259 c
  mirabilis Sandn. 38-693   Conn. N.C. S.C.
  antennalis Sandn. 38-695   Fla. Ala. Miss.
  quadrimaculata Horn 70-40   Que.-D.C.-
    sulcatus Blatch. 10-681 c   [Mich.
    blatchleyi Musg. 33-57 c
  musgravei Sandn. 38-698   N.Y.-Va.-Mo.
  sinuata Lec. 52-44   Ga.-S.C.-Miss.
  decorata Sandn. 38-701   D.C. Ind. Kan.
  vittipennis Zimm. 69-259   Que.-S.C.-Kan.
  convexula Sandn. 38-704   Fla.
  märkelii Mots. 54-12   Mass.-Ark.
  ?canaliculata Gyll. 08-552 c   Eur.
    elongata Mots. 59-51 c   ?N.A.

Simsonia Carter & Zeck 29-58 [1] [3]
  * vittata Melsh. 44-99 [3]   N.Y. Ind. Man.
  bivittata Lec. 52-44 [3]   Wis.
  quadrinotata Say 25-187 [3]   M.St. Can.
  brunnescens Fall 25-177 [3]   Cal.
  dietrichi Musg. 33-54 [3]   Fla. Ga. Miss.

c Revision of genus, Sanderson—38.

Elmis [1]
  immunis Fall 25-178
    cryophilus Musg. 32-79 [7]
  addenda Fall 07-226 [8]
  ornata Schffr. 11-120 [8]

Narpus Csy. 93-582 [1] [9]
  * concolor Lec. 81-75 [9]   B.C. Alb. Wash.
    solutus Brown 33-46 [7] [9]
  angustus Csy. 93-583 [9]   Cal.
    angustatus Hntn. 36-57
  arizonica Brown 30-90 [9]   Ariz.

Heterlimnius Hntn. 35-178 [1]
  * kœbelei Martin 27-68 [10]   Wash.
  divergens Lec. 74-52 [10]   B.C. Cal.
  tardellus Fall 25-179 [10]   Mass.
  cryophilus Musg. 32-79 [10]   Tenn.
  subarcticus Brown 30-241 [10] [11]   Que.
  trivittatus Brown 30-91 [10]   Que.
  ovalis Lec. 63-74 [10]   Pa. Ind.
  elegans Lec. 52-43 [10]   Vt.
  quadrimaculatus Horn 70-37 [10]   Cal.

Microcyllœpus Hntn. 35-178 [1] [10]
  * pusillus Lec. 52-44 [10]   M.St. Can.
    s.apta Musg. 33-56   Va. Fla.
    . *s.perdita* Musg. 33-56   Fla.
    s.lödingi Musg. 33-56   Ala. Miss.
  foveatus Lec. 74-53 [10]   Cal.
  similis Horn 70-38 [10]   Ariz.

Neoelmis Musg. 35-34 [1] [12]
  * cæsa Lec. 74-53 [12]   Tex.

Limnius Er. 47-522 [1] [13]
  * latiusculus Lec. 66-380 [13]   Pa.

Heterelmis [1]
  browni Hatch 38-16   Mont.

ANCYRONYCHINI [1]

Macronychus Müll. 06-207 [1] [13]
  * glabratus Say 25-187 [13]   E.U.S.
  thermæ Hatch 38-18   Mont.

Ancyronyx [1]

Zaitzevia Champ. 23-170 [1] [13]
  * parvulus Horn 70-41 [13]   B.C. Cal.
    columbiensis Angell 92-84 [14]

[7] Fall—34.
[8] Cited as probably the only American species remaining in the genus, Hinton—35.
[9] Review of genus, transferred from Dryopini, Hinton—36.
[10] Revision of genus, Hinton—35.
[11] As synonym of tardellus, Fall—34.
[12] Revision of genus, Musgrave—35.
[13] Revision of genus, Hinton—36.
[14] Sanderson—38.

HETEROCERIDÆ

Heterocerus
   LITTORIMUS Gozis 85-120
   compactus Fall 37-30   Man. Ia.

DASCILLIDÆ

Macropogon Mots. 45-38 [15]
* sequoiæ Hopping 36-46 [15]   Cal.
  testaceipennis Mots. 59-362 [15] Cal.-B.C.
   rubricollis Pic 27-34 [15]
   cribricollis Brown 29-274 [15] [16]

piceus Lec. 61-362 [15]   N.B.-Ill.-B.C.
  rufipes Horn 80-79 [15]
  dubius Brown 29-273 [15] [16]

Anorus Lec. 59-86 [17]
* piceus Lec. 59-87 [17]   Cal. L.Cal.
  parvicollis Horn 94-365 [17]   Ariz.
  arizonicus Blais. 34-323   Ariz.

CHELONARIIDÆ

Chelonarium Fahr. 01-101 [18]
* lecontei Thoms. 67-84 [18]   Fla.

NITIDULIDÆ

Brachypterolus Grouv. 13-387
* pulicarius Linn. 58-357 [19]   Eur. Wis.
                  [N.Eng.
Carpophilus
  humeralis Fab. 98-74   Cal. Fla.
   rickseckeri Fall 10-124 [20]

Nitidula
  carnaria Schall. 83-257 [19] Eur. Wis. Cal.

Phenolia Er. 43-299 [21]
* grossa Fabr. 01-347 [21]   Can.-Me.-Wyo.-
                  [Tex.

Soronia Er. 43-277 [21]
* guttulata Lec. 63-64 [21]   N.Y.-Ore.-Ariz.

Lobiopa Er. 43-291 [21]
* setosa Harold 68-104 [21]   Mass.-Ut.-B.C.
   setulosa || Lec. 63-63 [21]
   substriata Hamilt. 93-306 [21]
  oblonga Parsons 39-159   Cal.
  undulata Say 25-179 [21] Me.-Fla.-Mich.-
                  [Mex.

falli Parsons 39-161   Ariz.
  brunnescens Blatch. 17-238 [21] Fla. Mass.
  punctata Parsons 39-163   Fla. Jamaica
  ?insularis Cast. 40-10 [21]   Gulf.St.

Amphotis Er. 43-290 [21]
* ulkei Lec. 66-376 [21]   Mass.-D.C.
  schwarzi Ulke 87-77 [21]   Va. Ala. N.C.

Pocadius Er. 43-318 [22]
* fulvipennis Er. 43-319 [22]   Mex. Cal.
   dorsalis Horn 79-311 [22]
  helvolus Er. 43-320 [22] Conn.-Ga.-B.C.-
   breviusculus Reitt. 76-318 [22]   [Tex.
   ferrugineus Chevr. 63-604 [22]   [Mex.
   infuscatus Reitt. 74-94 [22]   [W.I.
   limbatus Reitt. 74-95 [22]
  niger Parsons 36-116   N.M. Ariz.
   fulvipennis || Fall & Ckll. 07-175 [22]
  basalis Schffr. 11-117 [22]   Ariz.

Cryptarcha Shuck. 39-165 [23]
   CRYPTARCHA (s.str.) [23]
  ampla Er. 43-356 [23] Que.-Fla.-Cal.-Or.
  (*grandicollis* Reitt., not No. Amer.)[23]
  glabra Schffr. 09-375 [23]   Ariz.
  strigatula Parsons 38-98   Mass.-Ga.-
                [Mich. Tex.
   strigata of authors (not Fabr.)
   concinna || Reitt. 73-150 [23]
   LEPIARCHA Shp. 91-385 [23]
   (*Cryptarchula* Gglb. 99-551)[23]
  gila Parsons 38-99   Ariz. Cal.
  concinna Melsh. 53-41 [23]   Mass.-Fla.-
   liturata Lec. 63-30 [23]   [Cal.-Ore.
   picta Melsh. 46-107 [23]
   bella Reitt. 73-150 [23]

HEMIPEPLIDÆ

Hemipeplus [24]

EROTYLIDÆ

Ischyrus
  quadripunctatus Cliv. 91-437
   alabamæ Schffr. 31-175 [25]
  a.antedivisa Mader 38-19   (no. loc.)

CRYPTOPHAGIDÆ

Cryptophagus
  dentatus Hhst. 93-15 [26] [27]   Arctic N.A.
  blumi Blais. 37-158   Mont.

---

[15] Revision of genus, R.Hopping—36.
[16] Fall—34.
[17] Revision of genus, Blaisdell—34.
[18] Placed in Dryopidæ by Mequignon—34.
[19] Dodge—37.
[20] Parsons in litt.
[21] Revision of group of genera, Parsons—39.

[22] Revision of genus, Parsons—36.
[23] Revision of genus, Parsons—38.
[24] Transferred from Cucujidæ (and placed near Melandryidæ and Oedemeridæ) by H. Scott—33.
[25] Mader—38.
[26] Brown—37.
[27] Blair—33.

COLYDIIDÆ

RHAGODERINI [28]

ORTHOCERINI

Rhagodera [29]

CORTICINI

Anchomma [29]

Phlœonemus
(*adhærens* Shp., not No. American) [29]

Microsicus [30]
parvulus Guer. 29-189 [30]   S.U.S. Mex.
setosus Shp. 94-456 [30]

Lasconotus Er. 46-258 [31]
(*Illestus* Pascœ 63-33) [29]
(*Lado* Wankow. 67-249) [29]
(*Othismopteryx* Sahlb. 71-44) [29]

Aulonium
bidentatum Fab. 01-556 [32]   L.Cal. C.A.
[W.I.

Lapethus Csy. 90-317 [33]
(*Lytopeplus* Shp. 95-494)
(*Brachylon* Gorh. 98-256)

LATHRIDIIDÆ

Corticaria
linearis Payk. 98-302 [34]   Eur. Sib.
[Greenl.

Enicmus
tricarinatus Brown 34-22   Sask.

MYCETÆIDÆ

Stethorhanis
borealis Blais. 34-325   B.C.

COCCINELLIDÆ

Hyperaspis
leachi Nunen. 34-19   Cal.
biornatus Nunen. 34-18   Cal.
fimbriolata Melsh. 46-180
s.marginatus Gaines 33-263   Tex.

Hyperaspidius
horni Nunen. 34-19   N.J.

Scymnus
maderi Nunen. 37-183   Cal.
quercus || Nunen. 34-18
scotti Nunen. 34-17   Cal.
schuberti Nunen. 34-17   Ariz. Cal.

Coccidula
suturalis Weise 95-132 [35]   Vanc.-N.J.
occidentalis Horn 95-114 [35]

Psyllobora
plagiata Schffr. 08-125   Ariz.
kœbelei Nunen. 11-71 [36]

Ceratomegilla
cottlei Nunen. 34-20   Wyo.

Adonia
amœna Fald. 35-453 [37]   Alaska, Asia

Hippodamia
hoppingi Nunen. 34-21   Cal.
lunatomaculata Mots. 45-382
a.lengi Joh. 10-865 [36]   Cal.
nigromaculata Nunen. 34-20 (Ad-
[alia) [36]

Cleis
concolor Crotch 74-142 [38]   Mex. ?Tex.

Anatis Muls. 51-133 [39]
* ocellata Linn. 58-366 [39]   Eur. Sib.
s.halonis Lew. 96-28 [39]
s.mali Say 24-93 [39]   N.H.-N.J.-Wash.
labiculata Say 35-288 [39]
rathvoni Lec. 52-132 [39]   Cal. Ore. Ida.
s.lecontei Csy. 99-98 [39]   Ariz. N.M.
[Colo. Wis.
quindecimpunctata Cliv. 08-1027 [39]
[Can.-N.C.-Ark.-Ill.
caseyi Westc. 12-422 [39]
signaticollis Muls. 50-134 [39]
[Ill.

Exochomus
californicus Csy. 99-107 [40]   Cal. Nev.
quatuorpustulatus Linn. 58-367 [40]   Eur.
[Cal.
quadripustulatus Nunen. 34-113

[28] This tribe unnecessary, the genera placed elsewhere, Hinton—35.
[29] Hinton—35.
[30] Hinton—36.
[31] This name was not validated by Erichson in
1845 (1846). Hinton—35 credits it to Leconte 59-
282 but Lacordaire was earlier.
[32] Fall—34.
[33] Placed in Murmidiidæ by Leng but in Colydiidæ by Hinton—36.
[34] Carpenter—38.

[35] Dodge—38.
[36] Nunenmacher in litt.
[37] Scott—33.
[38] Gaines—33.
[39] Revision of genus, McKenzie—36.
[40] Valid species, Nunenmacher—34.

Brumus
  blumi Nunen. 34-114   Cal.

Epilachna
  varivestis Muis. 50-815   N.A.
    corrupta Muis. 50-815 [41]
    a.genuina Muls. 50-817 [36]
    a.juncta Joh. 10-79 [36]

ALLECULIDÆ

Pseudocistela Cr. 73-108 [42]
  * (*Cistela* of Lec. & Horn) [42]
    (*Chromatia* Lec. 61-244) [42]
    amœna Say 23-268   Ind. C. Tex.
    brevis Say 23-269   E.Can. E.U.S.
      erythroptera Ziegl. 44-46
      rufipes Melsh.
    pinguis Lec. 59-16   N.M. Kan. Cal.
    pacifica R.Hopp. 33-284   B.C. Cal.
    opaca Lec. 59-78   Cal.
      theveneti Horn 75-156
    marginata Ziegl. 44-46
    pectinata R.Hopp. 33-285   B.C.

TENEBRIONIDÆ [1]

AUCHMOBIINI *

Auchmobius Lec. 51-139 [2]
  * sublævis Lec. 51-140 [2]   Cal.
    subovalis Blais. 34-238   B.C.
    slevini Blais. 34-243   Cal.
    parvicollis Blais. 34-246   Cal.
    angelicus Blais. 34-249   Cal.
    picipes Blais. 34-252   Cal.
    subboreus Blais. 34-254   Cal.
    sanfordi Blais. 34-257   Cal.
                  • • •

Stibia Horn 70-260 [3]
    (*Eutriorophus* Csy. 24-296 [3] [4]
    granulata Blais. 23-238   L.Cal.
    fallaciosa Blais. 36-70   L.Cal.
      puncticollis of Blais. (not Horn)[3]
      v.interstitialis Blais. 36-73   L.Cal.
    cribrata Blais. 23-239   L.Cal.
      opaca Blais. 25-329 (nom.nud.)[3]
    williamsi Blais. 25-328   L.Cal.
    puncticollis Horn 70-260 [3] Cal. L.Cal.
      hannai Blais. 25-329 [5]
    s.martinensis Blais. 36-83 L.Cal.

[36] Nunenmacher in litt.
[41] Brown—86.
[42] R.Hopping—33.
[1] Blaisdell—39. (Contains discussions of relationships and components of certain subfamilies and tribes arranged in two sections : (1) Tentyriinæ, Coniontinæ, Asidinæ (Craniotini, Asidini) ; (2) Eleodinæ (Eleodini, Amphidorini), Helopinæ (Stenotrichini, Helopini), Blaptinæ, Zopherinæ, etc., etc.)
[2] Revision of genus, Blaisdell—34.
[3] Revision of genus, Blaisdell—36.
[4] Blaisdell—33.

sparsa Blais. 23-237   L.Cal.
blairi Blais. 36-888   "Cal.merid."
tuckeri Csy. 24-297 [3] [4]   Ariz.
imperialis Blais. 36-94   Cal. Ariz.
tanneri Blais. 36-97   Cal.
tortugensis Blais. 36-100   L.Cal.

Eschatomoxys Blais. 35-125
  wagneri Blais. 35-125   Cal.

Craniotis [5]

Euschides   Cal.
  lecontella Blais. 36-227   Cal.
    v.tempestalis Blais. 36-229 Cal.
  speculatus Blais. 36-225   Cal.
  cressoni Blais. 33-191   Cal.

Eleodes
  paradoxa Blais. 31-78 [5]   Cal.
    montanus Blais. 25-385 [5]
      MELANELEODES Blais. 09-33
  lineata Blais. 39-55   Ariz.
  omissa Lec. 58-186
    s.tumida Blais. 33-194 [4] Cal.

                  • • •

spoliata Blais. 33-196   Ore.
acuta Say. 23-258
  s.pernigra Blais. 37-128   Tex.
    ELEODES (s.str.)
  dentipes Esch. 29-10
    s.sordida Blais. 35-30   Cal.

                  • • •

amedeensis Blais. 33-199   Çal. Nev.
armata Lec. 51-134
  v.pumila Blais. 33-197   Cal.
    ARPELEODES Blais. 37-128
  tibialis Blais. 09-311 [7]   L.Cal.

                  • • •

kaweana Blais. 33-203   Cal.
scabriventris Blais. 33-202   Cal.
oblonga Blais. 33-206   Cal.
  BLAPYLIS Horn 70-301
indentata Blais. 35-28   Wash.
  STENELEODES Blais. 09-33
ornatipennis Blais. 37-129   N.Mex.
  HOLELEODES Blais. 37-132
beameri Blais. 37-132   Ariz.
bryanti Blais. 37-134   Ariz.
palmerleensis Blais. 37-136   Ariz.

Neobaphion
  elongatum Blais. 33-208   Nev.

Eleodopsis Blais. 39-52 [8]
  subvestita Blais. 39-53 Sn Nich.Id.,Cal.

[5] Belongs in Asidinæ—Craniotini, before Asidini, Blaisdell—37.
[6] Blaisdell—35.
[7] Blaisdell—37.
[8] Belongs in Eleodopsinæ—Eleodopsini, to follow Eleodinæ.

[9] Delete, transfer genera to Amphidorini and Eleodini, Blaisdell——39.
[10] Results derived from rearing, Blaisdell—35.
[11] Revision of genus, Good—36.
[12] Other synonyms do not belong here, Good—36.
[13] Shepard in litt, Uyttenboogaart, etc.
[14] Blaisdell—34.
[15] Described in Hypophlœus.
[16] Revision of genus, Blaisdell—33.
[17] Valid species, Blaisdell—33.
[18] Transfer to Helopinæ—Stenotrichini, to follow Helops, Blaisdell—39.
[19] Blaisdell—33.

[20] Revision of family, Borchmann—36.
[21] Leng Catalog.
[22] Mank—37.
[23] Revision of genus, Mank—38.
[24] Hatch—34.
[25] In Anobiini near Gastrallus.
[26] Snyder—35.

Hadrobregmus
destructor Fisher 38-26   Alaska
 •
Platybregmus Fisher 34-275 [27]
canadensis Fisher 34-275   Ont.

Eupactus Lec. 61-203 [28]
(*Thaptor* Gorh. 83-205) [29]
oblongus Gorh. 83-206 [25]   Cal. Mex.

BOSTRYCHIDÆ

Endecatomus Mellie 47-108 [30][31]
* reticulatus Hbst. 93-70[31] Eur. S.St. Ind.
  rugosus Dej. MS (Gemm. & Har.)[31]
  rugosus Rand. 38-26 [31]   Me.-Va.-Tex.-
  dorsalis Mellie 48-218 [31]   [Mich.

Sinoxylon
simplex Horn 85-155   Tex.
brevicollis Csy. 98-70 (Amphice-
                              [rus) [32]

Rhizopertha
dominica Fahr. 92-359 [33]   Cosmop.
pusilla Fahr. 98-156 [33]
fissicornis Marsh. 02-82 [33]
piceus Marsh. 02-88 [33]
rufa Hope 45-16 [33]
frumentarius Nördl. 55-? [33]
moderatus Walk. 59-260 [33]

Heterarthron
(*parvulum* Lesne, not No. American) [34]

LYCTIDÆ

Trogoxylon [35]
caseyi Lesne 37-240   Tex.
rectangulus || Csy. 24-184 [34]

CISIDÆ

Octotemnus
dixiensis Tanner 34-47   Utah

[27] In Anobiini near Hadrobregmus.
[28] Van Dyke—36.
[29] Used also as subgenus, Van Dyke—36.
[30] Spelled Hendecatomus by Lesne and others.
[31] Revision of genus, Lesne—35.
[32] Lesne—37.
[33] Potter—35.
[34] Lesne—37.
[35] Placed in Bostrychidæ by Lesne—37.

SCARABÆIDÆ

Pinotus
(*spadiceus* Lued., *not No. American*) [1]

Phanæus
floridanus Dols. 24-94 [2]   Fla.
niger Dols. 24-95 [2]   La.
quadridens
  v.borealis Dols. 24-100 [1]
vindex
  v.cyanellus Rob. 38-107   Fla.

Onthophagus
depressus Har. 71-116
  carteri Blackb. 04-147 [3] Ga. S.Afr.

Aegialia
  LEPTÆGIALIA Brown 31-12
browni Saylor 34-74   Cal.
            • • •
conferta Horn 71-294 [4]   Ga. Ill. Cal.
  nigrella Brown 13-47 [4]   [Wash.

Trichlorhyssemus
* alternatus Hntn. 38-127 [5] Mex. ?Ariz.

Aphodius
utahensis Rob. 38-107   Utah
essigi Saylor 35-134   Cal.
rehni Rob. 38-109   Colo. Ariz.
linsleyi VanD. 33-115   Cal.
browni Hntn. 34-277
  smithi || Brown 30-2 [6]
rugoclypeus Hntn. 34-219   Cal.
  cadaverinus of Saylor 33-188 [6]
cadaverinus Mannh. 43-261 [6] Cal.
martini VanD. 28-153 [6]   Cal.
washtucna Rob. 38-108   Wash.
dilaticollis Saylor 35-80   Ore.

Euparia Serv. 25-357 [7]
* castanea Serv. 25-357 [7] Fla.-La. C.A.

Euparixia Brown 27-288 [1]
  (*Euparixa* Leng & Mutch. 33-39) [1]

Atænius
miamii Cartw. 34-200   Fla.
platensis Blanch. 37-185 [8] S.A. W.I.
  integer Har. 68-96 [8]   [N.C. Tex.
  anticus Fall 30-105 [8]
falli Hntn. 34-119
  consors || Fall 30-104 [6]
carolinus VanD. 28-157 [9]

Xeropsamobeus Saylor 37-36
* desertus VanD. 18-10   Cal.

[1] Chapin in litt.
[2] Transferred from Copris, Chapin in litt.
[3] Cartwright—38.
[4] Fall in litt.
[5] Hinton—38.
[6] Hinton—34.
[7] Revision of genus, Hinton—36.
[8] Hinton—37.
[9] Transferred from Aphodius, Hinton—37.

Pleurophorus
batesi Arrow [10]   W.I. C.A. N.A.
  parvulus of authors [10]
  nanus of Horn 87-96 [10]

Saprosites Redt. 58-436
* ventralis Horn 87-92 [11] Can. D.C. Ind.

Phæochrous Cast. 40-108 [12] [13]
*   (*Phæocrous* Pering. 08-646) [12]
  (*Silphodes* Westw. 45-160) [12]
emarginatus Cast. 40-109 [12'] E.Ind. ?Cal.
  behrensii Horn 67-163 [12]

Geotrupes
  (*lævistriatus* Mots., not *No. Amer.*) [14]
  (*occidentalis* Horn 80-144, *syn.*) [14]

Pleocoma Lec. 56-24 [15] [16]
* remota Davis 34-23 [15] [16] Utah
  behrensi of Smith 85-33 [16]
shastensis VanD. 33-183 [15] [16] Cal.
puncticollis Rivers 89-17 [15] [16] Cal.
rickseckeri Horn 88-5 [15] [16] Cal.
trifoliata Linsley 38-57 ?Alaska
australis Fall 11-65 [15] [16] Cal.
bicolor Linsley 35-11 [15] [16] Cal.
carinata Linsley 38-56 Ore.
simi Davis 34-24 [15] [16] Ore.
oregonensis Leach 33-186 [15] [16] Ore.
fimbriata Lec. 56-25 [15] [16] Cal.
tularensis Leach 33-186 [15] [16] Cal.
  fimbriata of Davis 35-17 [16]
behrensi Lec. 74-83 [15] [16] Cal.
sonomæ Linsley 35-12 [15] [16] Cal.
hirticollis Schauf. 70-58 [15] [16] Cal.
  fimbriata of Lec. 57-40 [16]
s.vandykei Linsley 38-56 Cal.
crinita Linsley 38-53 Cal.
hoppingi Fall 06-394 [15] [16] Cal.
badia Fall 17-15 [15] [16] Cal.
conjungens Horn 88-7 [15] [16] Cal.
hirsuta Davis 34-88 [15] [16] [17] Cal.
blaisdelli Linsley 38-55 Cal.
  conjungens of Leach 33-185 [16]
staff Schauf. 70-52 [15] [16] Cal. [18]
  adjuvans Crotch 74-58 [16]
dubitalis Davis 35-30 [15] [16] [19] Ore.
s.leachi Linsley 38-52 Ore.
minor Linsley 38-52 Ore.
edwardsi Lec. 74-83 [15] [16] Cal.
  staff of Horn 88-11 [16]
  ulkei of Leach 33-184 [16]
ulkei Horn 88-9 [15] [16] ?Utah, [20] Cal.

[10] Cartwright in litt.
[11] Transferred from Pleurophorus, Chapin and Cartwright in litt.
[12] Robinson—38.
[13] To follow Pachyplectris.
[14] Robinson—38, and Chapin in litt.
[15] Revision of genus, A. C. Davis—35.
[16] Linsley—38; a more complete revision than that of Davis, but arranged alphabetically.
[17] Described as a variety of conjungens but elevated to species by Linsley—38.
[18] This probably should be changed to Oregon, Leach—33.
[19] Described as a variety of staff but elevated by Linsley—38 and locality cited as California.
[20] Found only in California, Leach—38.

Benedictia Sanderson 39-1 [21]
* pilosa Sandn. 39-2 Tex.

Lichnanthe Burm. 44-26
  (*Amphicoma* of authors) [22]

Oncerinæ [23]

Oncerus Lec. 56-283 [23]
* floralis Lec. 56-284 [23] Cal.

Nefoncerus Saylor 38-101 [23]
* convergens Horn 94-394 [23] Cal.

* * *

Chaunocolus Saylor 37-35 [24]
* cornutus Saylor 37-35 L.Cal.

Chnaunanthus Burm. 44-31 [25]
*   (*Acratus* Horn 67-165) [25]
  (*Pseudacratus* Dalla Torre 12-7) [25]
flavipennis Horn 67-165 [25] Ariz. Ut.
  palmeri Horn 94-393 [25]
discolor Burm. 44-32 [25] Mex. L.Cal.
chapini Saylor 37-535 [25] Cal. Ore.

Serica MacLeay 19-146 [26]
  (*Aserica* Lewis 95-394) [26]
porcula Csy. [27]
alternata Lec. 56-276 N.M.-Cal. Colo.
                               [Ut.
s.exolita Dawson 33-437 Cal.
s.patruela Dawson 33-437 Cal.
acontia Dawson 33-438 Cal.
repanda Dawson 33-439 Cal.
laguna Saylor 35-1 Cal.
peregrina Chpn. 38-68 Japan, N.Y.
  similis of Leng & Mutch. 33-90 [28]
  brunnea || Waterh. 75-101 [28]
mckenziei Saylor 35-2 Cal.
prunipennis Saylor 36-4 Cal.
  pruinosa || Saylor 35-2 [29]
falcata Daws. 33-439 Nev. Cal.-Wash.
stygia Daws. 33-439 Cal.
prava Daws. 33-440 Cal.
senta Daws. 33-440 Cal.

Autoserica Brenske 97-356 [26]
*   (*Aserica* of Arrow, not Lewis) [26]
castanea Arrow 13-398 Japan,N.Y.N.J.

Diplotaxis
falli Saylor 35-35 Cal.
testacea Burm. 55-263 [30] Carolina
basalis Fall 09-73 Kan. Neb. Colo.

[21] Belongs in Pleocominæ, Sanderson—39.
[22] Chapin—38.
[23] Revision of subfamily, Saylor—38.
[24] Probably in Chasmatopterini of Leng Catalog.
[25] Revision of genus, Saylor—37.
[26] These names are arranged thus by Chapin—32, but Autoserica is placed as synonym of Aserica Lewis by Arrow—27 and —33.
[27] Valid species, Dawson in litt.
[28] Chapin—38.
[29] Saylor—38.
[30] These references were reversed in the catalog; the species are probably distinct, Fall in litt.

[31] Travis—34.
[32] Saylor—37.
[33] Used as subgenus by Blocker—36.
[34] Also used as subgenus by Saylor in litt,. and other writers.
[35] Saylor in litt.
[36] Von Bloeker—37.
[37] Sanderson—39.
[38] Saylor—35.
[39] Fall—37.

d Revision of subgenus, Sanderson—39.
[1] Chapin in litt.
[12] Robinson—38.

Chlænobia Blanch. 50-116 [40]
* vexata Horn 85-120 (Phytalus)[40]   Tex.
    cavifrons Linell 96-729

Polyphylla
  conspersa Burm. 55-407 [41]
    r.hammondi Lec. 56-228 [41]   Kan.-Ill.
    r.subvittata Lec. 56-229 [41]   Tex.
      squamicauda Csy. 14-324 [41]
      oblita Csy. 14-326 [41]
      bisinuata Csy. 14-327 [41]
    r.molesta Csy. 14-324 [41]   N.Mex.
      ?verecunda Csy. 14-325 [41]   N.Mex.
      ?impigra Csy. 14-326 [41]   N.Mex.
      ?sejuncta Csy. 14-328 [41]   N.Mex.
    r.proba Csy. 14-329 [4]   Ariz.
      ?diffusa Csy. 14-329[1]   Ariz.
      ?pimalis Csy. 14-330 [41]   Ariz.
      (*r.cavifrons Lec., not No. Amer.*)[41]
    barbata Cazier 38-161   Cal.
    hirsuta VanD. 33-116   Ariz.
    occidentalis Linn. 67-555 [41]
    r.variolosa Hentz 30-256 [41]   Me.-N.Y.
    r.occidentalis (s.str.)[41]   Va.-Fla.
    r.gracilis Horn 81-73 [41]   Fla.
    r.comes Csy. 14-352 [51]   Ky.
    r.speciosa Csy. 89-17 [41]   N.M. Colo.
      10-lineata of Csy. (part.)[41]
      acomana Csy. 14-342 [41]
      latifrons Csy. 14-340 [41]
      ?diffracta Csy. 89-17[41]   N.Mex.
      ?adusta Csy. 14-331[41]   N.Mex.
    r.10-lineata Say 23-246 [41]   Kan.
      ?reducta Csy. 14-346[41]   Wash.
      ?nigra Csy. 14-334[41]   Wash.
      ?crinita of Csy.[41]   Wash.
    r.perversa Csy. 14-348[41]   Wash.
    r.oregona Csy. 14-348 [41]   Ore.
      ?opposita Csy. 14-330 [41]   Ore.
      ?mystica Csy. 14-334 [41]   Ore.
      ?crinita of Csy. (part)[41]
    r.modulata Csy. 14-333 [41]   Ore.
    r.arguta Csy. 14-339 [41]   Utah
    r.laticauda Csy. 14-345 [41]   Nev.
    r.rugosipennis Csy. 14-337 [41] Ariz. Ut.
    r.matrona Csy. 14-350 [41]   Ariz.
    r.fuscula Fall 08-161 [41]   Ariz.
      ?lævicauda Csy. 14-338 [41]   Ariz.
      ?r.crinita Lec. 56-230 [41]   Cal.
      ?r.incolumis Csy. 14-335 [41]   Cal.
      ?r.relicta Csy. 14-336 [41]   Cal.
      ?r.robustula Csy. 14-336 [41]   Cal.
      ?r.sobrina Csy. 14-339 [41]   Cal.
      ?r.squamotecta Csy. 14-343 [41]   Cal.
      ?r.pacifica Csy. 95-607 [41]   Cal.
      ?r.ruficollis Csy. 14-346 [41]   Cal.
      ?r.castanea Csy. 14-347 [41]   Cal.

Phobetus Lec. 56-227 [42]
* comatus Lec. 56-227 [42] [43]   Cal. Ore.
    leachi Barrett 35-51 [42]
    v.centralis Csy. 09-281 [42] [44]   Cal.
    v.sloopi Barrett 33-130 [42] [43]   Cal.

saylori Cazier 37-116   Catal.Id.Cal.
ciliatus Barrett 35-51 [42] [43]   Cal.
testaceus Lec. 61-346 [42] [43]   Cal.
  collacatus Walker 66-346 *
humeralis Cazier 37-85   Cal.
mojavus Barrett 33-130 [42] [43]   Cal.
palpalis Saylor 36-1 [42]   Cal.

Dichelonyx
  arizonica Barrett 33-132   Ariz.
  vicina Fall 01-291
    deserta Hopping 31-236 [35]
  arizonensis Saylor 33-158   Ariz.

Cœnonycha
  tingi Cazier 37-126   Cal.
  testacea Cazier 37-127   Cal.
  stohleri Saylor 35-102   Nev.

Plectris Serv. 25-369
*   (*?Philochlœnia* Blanch. 50-122)[45]
  aliena Chpn. 34-34   S.Car.

Ceraspis Serv. 25-370
*   (*Faula* Blanch. 50-124)[46]
  pilatei Har. 63-174 [35]   Ariz. C.A.
    baui Nonfr. 90-76 [46]

Leptohoplia Saylor 35-132 [47]
* testaceipennis Saylor 35-133   Cal.

Hoplia
  lecontei DallaT. 13-376   Cal.
    pubicollis || Lec. 56-285 [38]
  floridana, for Ind. read Fla.[1]

Anomala
  foraminosa Bates 88-229 [12]   Mex. Tex.
  insitiva Rob. 38-112   Tex.
    polychalca of Schffr. 06-2 [12]
  dubia Scop. 63-3 [e]   Eur. N.J.

Rutelinæ [48]

Areodina

Cotalpa Burm. 44-423 [48]
*   COTALPA (s.str.)[48]
  consobrina Horn 71-337   Ariz.
  flavida Horn 78-53   Ut. Ariz.
  lanigera Linn. 64-22   Can.-N.J.
    s.obesa Csy. 15-90 [48]   Ind. Ia. Wis.
  molaris Csy. 15-90   La.
  subcribrata Wickh. 05-3   Kans.
  tau Wickh. 05-2   Ariz.
  vernicata Csy. 15-91   Va. N.Y.
    PARACOTALPA Chaus 15-256[48]
    (*Pocalta* Csy. 15-68) [48]
  brevis Csy. 15-95   Cal.
  granicollis Hald. 52-374   Utah
  lævicauda Csy. 15-95   Cal.

[40] Chapin—35.
[41] Kuntzen—33.
[42] Revision of genus, Cazier—37.
[43] Revision of genus, Barrett—35; all as valid species.
[44] As synonym, Barrett—33.

[e] Angell in litt.
[45] Chapin—34.
[46] Dalla Torre—Col. Cat. p. 59.
[47] Probably a new tribe near Hopliini, Saylor—35.
[48] Revision of subfamily, Ohaus—34.

nigripennis Csy. 15-97   Cal.
pubicollis Csy. 15-98   Utah
puncticollis Lec. 63-78   N.Mex.
rotunda Csy. 15-96   Cal.
rubripennis Csy. 15-97   Cal.
ursina Horn 67-168   Cal.

PARABYRSOPOLIS Chaus [15-256 [48]
    (Parareoda Csy. 15-68)[48]
arizonæ Chaus 12-313   Ariz.
batesi Chaus 15-257   Mex. ?Ariz.
    lanigera Bates 88-291 [48]
rufobrunnea Csy. 15-100   Ariz.
    Pelidnotina

Plusiotis Burm. 44-417 [48]
* woodi Horn 85-124   Tex.
ampliata Csy. 15-82   Ariz.
heyeri Skinner 05-289   Ariz. Mex.
    s.ocularis Csy. 15-83 [48]
angustata Csy. 15-86   Ariz.
lecontei Horn 82-120   Ariz. N.M. Mex.
gloriosa Lec. 54-221   Ariz. Mex.

Pelidnota MacLeay 19-157 [48]
    (Aglycoptera Shp. 85-2)[48]
    (Odontognathus Lap. 40-139)[48]
    (Strigidia Burm. 44-388)[48]
    PELIDNOTA (s.str.)
lucæ Lec. 63-78   L.Cal.
lugubris Lec. 74-54   Ariz. Mex.
lutea Cliv. 89-23   Tex. Fla.
s.brevicollis Csy. 15-74 [48]   Fla.
s.hudsonica Csy. 15-74 [48]   N.Y.
s.pallidipes Csy. 15-74 [48]   Va.
s.texensis Csy. 15-74 [48]   Tex.
oblonga Csy. 15-73   La.
s.debiliceps Csy. 15-73 [48]   N.J.
s.ponderella Csy. 15-73 [48]   D.C.
punctata Linn. 58-350   N.A. ?Mex.
tarsalis Csy. 15-74   N.Y.

    Parastasiina

Parastasia Westw. 41-204
*   (Barymorpha Guer. 43-41)[48]
    (Cœlidia Burm. 44-331)[48]
    (Echmatophorus Waterh. 95-158)[48]
    (Polymœchus Lec. 56-23)[48]
    (Urleta Westw. 75-238)[48]
brevipes Lec. 56-23   Pa. N.Y. Mo.
conicicollis Csy. 15-104   Pa.
    Rutelina

Rutela Latr. 02-151
*   (Microrutela Bates 04-250)[48]
formosa Burm. 44-383   Cuba, Fla.

Coscinocephalus Prell 36-145
    (Anoplocephalus || Schffr. 06-259)[49]

[48] Revision of subfamily, Ohaus—34.
[49] Prell—36.

Cyclocephala Latr. 29-552
*   (Diapatalia Csy. 15-111)[50]
    (Aclinidia Csy. 15-113)[50]
    SPILOSOTA Csy. 15-112 [51] [52]
    (Ochrosidia Csy. 15-112)[51] [52] [53]
hirta Lec. 61-346 [*2]   Cal. Ariz. Ut.
nigricollis Burm. 47-54 [52]
robusta Lec. 66-79 [52]
magister Csy. 15-132 [52]
palidissima Csy. 15-133 [52]
inconspicua Csy. 15-133 [52]
s.pilosicollis Saylor 36-2 [52]   Cal.
villosa Burm. 47-54 [52]   ?Cal.
abrupta Csy. 15-152 [52]   Cal. Ore. Ariz.
phasma Csy. 15-153 [52]
obesula Csy. 15-156 [52]
oblongula Csy. 15-156 [52]
rustica Csy. 15-157 [52]
reflexa Csy. 15-153 [52]
pasadenæ Csy. 15-148 [52]
arizonica Csy. 15-149 [52]
immaculata Cliv. 90-29 [52]   ?Cal.
nigrifrons Panz. 94-? [52]
frontalis Sturm. 43-116 [52]
rufifrons Csy. 15-145 [52]
    DICHROMINA Csy. 15-160 [51] [52]
dimidiata Burm. 47-57 [52]
    • • •
knobelæ Brown 34-33   Ark.
borealis Arrow 11-172   N.Y.
villosa Burm. 47-54 [25]
californica Arrow 37-9   Cal.
rustica Csy. 15-157 [54]

Ligyrodes Csy., belongs in Cyclocephalini after Dyscinetus [1] [55]

Cheiroplatys
    (Pseudaphonus Csy. 15-178)[54]
    (Orizabus Fairm. 78-260)[54]

Aphonus Lec. 56-21 [56]
* castaneus Melsh. 46-138 [56]   Me.-S.C.
obesus Burm. 47-119 [56]
trapezicollis Csy. 15-219 [56]
saginatus Csy. 15-220 [56]
cubiformis Csy. 15-221 [56]
densicauda Csy. 15-216 [56]   Pa.
tridentatus Say 23-209 [56]   Me.-Fla.-Ind.
frater Lec. 56-22 [56]
elongatus Csy. 15-220 [56]
aterrimus Csy. 15-216 [56]
congestus Csy. 15-218 [56]
politus Csy. 15-218 [56]
modulatus Csy. 15-219 [56]
scutellaris Csy. 24-335 [56]
variolosus Lec. 48-88 [56]   Ga. Fla.
hydropicus Lec. 56-22 [56]
ingens Csy. 24-334 [56]

[1] Chapin in litt.
[55] Saylor in litt.
[50] Arrow—37.
[51] Considered as a synonym of Cyclocephala by Arrow—37.
[52] Revision of subgenera Spilosota and Dichromina. Saylor—37.
[53] Used as valid genus by Brown—34.
[54] Arrow—37.
[55] Placed as synonym of Ligyrus by Arrow—37.
[56] Revision of genus, Arrow—37.

Aphonides Rivers 89-6
  (*Anoplognathus* || Rivers 89-101)[49]

Strategus Hope 37-87
  (*Anastrateyus* Csy. 15-231)[54]
  anteus Drury 73-74   Fla. Tex.
  atrolucens Csy. 15-247 [54]
  septentrionis Csy. 15-249 [54]
  divergens Csy. 15-246 [54]
  pinorum Csy. 15-248 [54]

Megasoma Kby. 25-566
  (*Megasominus* Csy. 15-261)[54]

Gymnetis MacLeay
  chevrolati G. & P.
  s.ramulosa Bates 69-389 [57] Nic. Ariz.
  ramifera Schffr. 05-159 [57]
  balteata Csy. 15-280 [57]

Cineretis Schürh. 37-56 [58]
  * argenteola Bates 89-354 [57]
  s.argenteola (s.str.)[57]   Mex.
  a.lætula Csy. 15-281[57]   Ariz.

Potosia Muls. 71-669 [59]
  * affinis Ander. 97-154   Eur. Asia, Cal.

Euphoria
  casselberryi Rob. 37-163   Tex.

Euphoriaspis Csy. 15-333
  * hirtipes Horn 80-401   Neb.
  æstuosa Horn 80-400 [60]   Kan.

CREMASTOCHEILINI [61]

Cremastocheilus Knoch 01-115 [61]
  (*Trinodia* Csy. 15-365)[61]
  (*Anatrinodia* Csy. 15-369)[61]
  lengi Cazier 38-86   Ariz.
  MACROPODINA Csy. 15-344 [61]

Genuchinus Westw. 74-23 [61]
  (*Psilocnemis* Burm. 42-676)[61]
Lissomelas Bates 89-376 [61]

Osmoderma
  * scabra Beauv.[62]
  eremicola Knoch [62]

[57] Schürhoff—87.
[58] To follow Gymnetis, Schürhoff—37.
[59] Intercepted by quarantine at San Francisco, Ting—34.
[60] Transferred from Euphoria, Chapin in litt.
[61] Generic revision of tribe, Cazier—38.
[62] These are the only nearctic species, Hoffmann in litt.

Trichiotinus Csy. 15-381 [63]
  * piger Fahr. 75-41[63] Me.-Fla.-Tex.-Minn.
  drummondi G. & P. 33-88 [63]
  rotundicollis Kby. 37-138 [63]
  lunulatus LeBaron 71-194 [63]
  reductus Csy. 15-384 [63]
  texanus of Leonard 28-430 [63]
  rufobrunneus Csy. 14-375 [63]   Fla.
  obesulus Csy. 14-376 [63]
  affinis of Blatch. 30-34 [63]
  texanus Horn 76-194 [63] Kan. Ckla. Tex.
  monticola Csy. 15-383 [63]   [N.M.
  intermedius Csy. 15-383[63]
  assimilis Kby. 37-137 [63] N.S.-Pa.-N.M.-
  bistriga Newm. 38-170 [63]   [B.C.
  piger of Burm. & Sch. 40-413 [63]
  variabilis Burm. & Sch. 41-240
            [(part)[63]
  bibens Burm. 42-756 (part)[63]
  affinis of Schaum 49-293 (part)[63]
  affinis G. & P. 33-93 [63] N.H.-Ga.-Ill.
  piger B. & S. 40-413 (part)[63]
  variabilis B. & S. 41-240 (part)[63]
  bibens Burm. 42-755 (part)[62]
  mutabilis Schaum 44-400 (part)[63]
  ventricosus Csy. 15-386 [63]
  parvulus Csy. 15-387 [63]
  viridans of Leonard 28-430 [63]
  viridans Kby. 37-139[63] Mich.-Kan.-Minn.
  piger B. & S. 40-413 (part)[63]
  variabilis B. & S. 41-240 (part)[63]
  bibens Burm. 42-755 (part)[63]
  affinis Schaum 49-293 (part)[63]
  lunulatus Fahr. 75-41 [63] Va.-Fla.-Tex.
  viridulus Fahr. 75-820 [63]
  virens Gmelin 90-1584 [63]
  piger of Schönh. 17-105 (part)[63]
  bibens Burm. 42-755 (part)[63]
  mutabilis Schaum 44-400 (part)[63]
  semiviridis Csy. 14-376 [63]
  carolinensis Csy. 14-376 [63]
  rasilicauda Csy. 15-389 [63]
  rufiventris Csy. 15-390 [63]
  bibens Fahr. 75-40 [63] N.Y.-Ga.-Ill.
  bidens Oliv. 89-62 [63]
  viridulus of Hubb. & Schw. 78-655[63]

LUCANIDÆ

Pseudolucanus
  capreolus Linn. 64-32
  a.nigricephalus Benesh 39-273   Ill.
  a.muticus Thunb. 06-205 [64] Md. Tenn.

Dorcus
  brevis Say 25-202 [65]   N.J.-Ga.-Mo.
  parallelus Say 24-248 [65]   Bor. Amer.
  parallelopipedus Linn. 58-354 e Eur. Que.

e Angell in Litt.
[63] Revision of genus, Hoffmann—35.
[64] Benesh—39.
[65] Benesh—37.

Platycerinæ

Platycerus [66]

PASSALIDÆ [7]

Pseudacanthinæ

Popilius Kaup 71-75
* (Passalus of authors)[67]
  disjunctus Ill. 00-78   N.A.-S.A.
  cornutus Fahr. 01-256 [67]
  interruptus Linn. 64-35 (part)[67]
  distinctus Weber 01-79 [67]

Passalinæ

Passalus Fahr. 92-24
* (Neleus || Kaup 69-30)[67]
  interruptus Linn. 58-354   L. Cal.
  sulcatus Scop. 68-76 [67]   [Tex.-S.A.
  spectabilis Perty 30-55 [67]
  grandis Dej. 37-194 [67]
  tlascala Perch. 35-45 [67]
  punctiger St. Farg. & Serv. 25-20
  synonyms [68]   [Tex.-S.A.

CERAMBYCIDÆ

Aplagiognathus Thoms. 60-320 [1]
* remotus Linsley 34-161   Ariz.

Stenodontes
  masticator Thoms. 67-99 [2]   Ariz. C.A.
                  [S.A.

Prionus
  californicus Mots. 45-89
  crassicornis Lec. 52-108 [3]
  ineptus Csy. 12-742 [3]
  humeralis Csy. 24-216 [3]
  s.horni Lam. 12-243 [4]   Ariz.
  lecontei Lam. 12-244 [4]   Cal.

Spondylis Fabr. 75-159
* upiformis Mannh. 43-304   Alas.-L.Sup.-
  laticeps Lec. 50-233 [3]   [Ariz.-Cal.
  collaris Csy. 12-218 [3]
  robustula Csy. 12-219 [3]
  subpubescens Csy. 12-219 [3]
  basalis Csy. 12-220 [3]
  parva Csy. 24-226 [3]

[66] Transferred from Dorcinæ, Benesh—37.
[67] Revision of family, Hincks & Dibb—35.
[68] See Col. Cat. pars 142 for list of 47 synonyms of this species.
[1] Belongs in subfamily Prioninæ, tribe Macrotomini, and subtribe Archetypi, Linsley—34.
[2] Subgenus Mallodon, Linsley—34.
[3] Linsley—38.
[4] Linsley—35.

Megasemum Kr. 79-97
  (Nothorhina || Csy. 12-263)[5]
  aspera Lec. 54-18 [5]   Ore. Vanc.

Tetropium
  auripile Bates 85-435 [6]   Mex. Ariz.

Opsimus Thoms. 60-377
* quadrilineatus Mannh. 43-305 Alas.-Cal.
  biplectralis Csy. 24-229 [3]

Oeme
  californica Linsley 34-162   Cal.
  laticollis Linsley 34-163   Cal.
  gracilis Lec. 81-27
  densicolle Csy. 24-250 (Paranop-
                     [lium)[6]

Pseudomethia Linsley 37-65
  arida Linsley 37-66   Cal.

Methia
  evaniformis Knull 37-306   Tex.
  xanthocollis Knull 35-98   Tex.
  juniperi Linsley 37-64   Cal.
  brevis Fall 29-58 [6]   L.Cal.

Osmidus Lec. 73-177
* guttatus Lec. 73-178   L.Cal. Ariz.
  obscurella Csy. 24-255 [3]
  vestitus Csy. 24-255 [3]

Brothylus
  gemmulatus Lec. 59-80   Cal. Utah
  consors Csy. 24-254 [3]
  longicollis Csy. 24-254 [3]

PHORACANTHINI [7]

Romaleum White 55-309 [7]
  (Thersalus Pascoe 66-372)[8]
  (Hypermallus Lac. 69-302)[8]

Eustromula Ckll. 06-242 [7]
* (Eustroma Lec. 73-186)[8]
  validum Lec. 58-82   Tex. Cal. L.Cal.
  huachucæ Csy. 24-245 (Anoplium)[8]

Elaphidion Serv. 34-66 [7]
  (Centrocerum Thoms. 64-244)[8]
  (Cyclopleurus Hope 35-107)[8]
  (Hypermallus || Csy. 12-292)[8]

Anelaphus Linsley 36-464
* spurcum Lec. 53-442 [9]   Tex.
  truncatum Hald. 47-33 [9]   Fla. Tex.
  niveivestitum Schffr. 05-132 [9]   Tex.
  brevidens Schffr. 08-333 [9]   Ariz.
  quadricollis Csy. 24-246 [9]   Ariz.
  inerme Newm. 40-29 [9]   Pa. Fla. Tex.
  albofasciatus Linell 96-393 [9]   Cal. Ariz.
  linelli Csy. 24-246 [9]

[5] Van Dyke—37.
[6] Linsley—34.
[7] Rearrangement of genera in the tribes Phoracanthini and Sphærionini, Linsley—36.
[8] Linsley—36.
[9] Transferred from Anoplium by Linsley—36.

Anopliomorpha Linsley 36-465 [7]
* rinconium Csy. 24-248 [9]   Ariz.
   reticollis Bates [9]   ?Ariz. L.Cal. Mex.

Anoplium Hald. 47-34 [7]
* nanulum Csy. 24-247 [8]   Ariz.
   tuckeri Csy. 24-247 [8]   Ariz. Tex.
   hoferi Knull 34-69 [8]   Ariz.
   simile Schffr. 08-334 [8]   Ariz.
   mœstum Lec. 53-442 [5]   Fla. Tex.
   duncani Knull 27-117 [8]   Ariz.
   unicolor Hald. 47-34 [8]   Pa. Tex.
     cinerascens Lec. 50-15 [8]
   magnipunctata Knull 34-12   N.Mex.
   tricallosum Knull 38-140   Ariz.
   nanum Fabr. 92-300 [8]   Cuba, Fla.
     subtropicum Csy. 24-245 [8]

Elaphidionopsis Linsley 36-467 [7]
* fasciatipennis Linsley 36-467   Tex.

SPHÆRIONINI [7]

Axestinus Lec. 73-177 [7]
*   (*Proteinidium* Bates 92-149) [8]
   obscurus Lec. 73-177   Tex.

Aneflus Lec. 75-185 [7]
* sonoranus Csy. 24-241 Ariz. N.M. Tex.
     obscurus of Leng 85-pl. II, f. 27. [8]
   prolixus Lec. 63-203   L.Cal.-Tex.
     fisheri Knull 34-335 [8]
   protensus Lec. 58-82   Cal.-Tex. Mex.
     cochisensis Csy. 12-296 [8]
   calvatus Horn 85-132   Ariz. Cal.

Aneflomorpha Csy. 12-291 [7]
* duncani Linsley 36-472   Ariz.
   parkeri Knull 36-334   Ariz.
   texana Linsley 36-473   Tex.
   elongata Linsley 36-473   Cal.
   californica Linsley 36-476   Cal.
   levetti Csy. 92-29 [10]   Ariz.
   arizonica Linsley 36-475   Ariz.
   fisheri Linsley 36-475   Tex.

Anepsyra Csy. 12-291 [7]
* tenue Lec. 54-81 [8]   Tex. Ariz.
   aculeatum Lec. 73-184 [11]   Tex.

Stenelaphus Linsley 36-477 [7]
* alienum Lec. 75-173   Ariz.

Stenosphenus Hald. 47-39 [7]
   notatum Cliv. 95-61   E.St.

Gymnopsyra Linsley 37-67 [7]
* phoracanthoides Linsley 37-68   Tex.

Psyrassa Pascoe 66-481 [7]
   (*Pseudibidion* Csy. 12-291) [7]
   brevicornis Linsley 34-164   Tex.
   basicornis Pasc. 66-481   Mex. Tex.

* * *

[10] Transferred from Aneflus by Linsley—36.
[11] Transferred from Aneflomorpha by Linsley—36.

Obrium
   glabrum Knull 37-41   Tex.

LEPTURINI [12]

Pyrotrichus Lec. 62-41 [12]
* vitticollis Lec. 62-41   Cal. B.C.
     cribripennis Csy. 13-198 [12]

Leptalia Lec. 73-204 [12]
* frankenhaeuseri Mannh. 53-252   Alas.-
     macilenta Mannh. 53-253 [12]   [Cal.
     fuscicollis Lec. 57-65 [12]

Encyclops Newm. 38-392 [12]
* cœrulea Say 27-280   Ont.-Conn.-Ill.
   pallipes Newm. 38-392 [12]
   californicus VanD. 20-45   Cal.

Toxotus Dej. 21-112 [12] [13]
*   (*Stenocorus* || Csy. 13-205) [12]
   cylindricollis Say 23-4   N.Y.-Ga.-Mo.
     dives Newm. 41-68 [12]
     dentipennis Hald. 47-58 [12]
     atratus Hald. 47-58 [12]
   flavolineatus Lec. 54-18   B.C. Wash.
                     [?Cal.
   vittiger Rand. 38-29   Que.-Pa.-Minn.
     nigripes Hald. 47-58 [12]
   pacificus Csy. 13-209   Cal.
     hesperus Csy. 13-210 [12]
     parviceps Csy. 13-210 [12]
     tenellus Csy. 13-211 [12]
   trivittatus Say 23-422   Man. Mo. Miss.
     virgatus Lec. 74-67 [12]
   schaumi Lec. 50-320   Que.-Vt.-Ill.-Minn.
   croceus Leng. 90-68 [12]
   cinnamopterus Rand. 38-45   Mass.-
                  [N.C.-Kan.
   lateralis Csy. 91-37   Cal.
     uteanus Csy. 24-273 [12]
     marginellus Csy. 24-274 [12]
   vestitus Hald. 47-59   B.C.-Cal.
     nubifer Lec. 59-80 [12]
     ater Leng 90-68 [12]
     truncatulus Csy. 13-211 [12]
     apiciventris Csy. 13-211 [12]
     flaccidus Csy. 13-212 [12]
     rufipennis Csy. 13-212 [12]
     plagiatus Csy. 24-273 [12]
     morio Csy. 24-274 [12]
   obtusus Lec. 73-206   B.C. Wash. Wyo.
     brevicollis Csy. 13-214 [12]   [Alb.
   aureatus Csy. 13-213   Cal.
     gilvicornis Csy. 13-213 [12]
     sericatus Csy. 13-213 [12]
     subpinguis Csy. 13-213 [12]
   oregonensis Csy. 13-214   Cal.-B.C.-Alb.

[12] Revision of part of tribe, Hopping—37.
[13] Linsley — 38 changes the authority of these genera to Zetterstedt 28-376) (74) ; however, they are amply validated by Dejean in 1821 (p. 112) and must be credited to him.

Stenocorus Geoff. 62-221 [12]
*   (*Rhagium* Fabr. 75-182) [12]
   inquisitor Linn. 58-393   Alas.-Cal.-Conn.
                  [Que. Eur. Sib. Japan
   indigator Fahr. 87-145 [12]
   lineatum Cliv. 95-13 [12]
   investigator Mannh. 52-367 [12]
   californicum Csy. 13-195 [13]
   crassipes Csy. 13-195 [12]
   parvicornis Csy. 13-195 [12]
   boreale Csy. 13-195 [12]
   cariniventre Csy. 13-196 [12]
   thoracicum Csy. 13-196 [12]
   montanum Csy. 13-197 [12]

Centrodera Lec. 50-325 [12]
*   (*Parapachyta* Csy. 13-216) [12]
   spurca Lec. 57-63   B.C.-Cal.-Nev.-Ida.
   cervinus Walk. 66-332 [12]
   decolorata Harris 41-93   Que.-N.Y.-Ill.
   rudibus Hald. 47-58 [12]
   lacustris Csy. 18-416 [12]
   nevadica Lec. 73-205   Cal. Nev.
   oculata Csy. 13-202 [12]
   blaisdelli VanD. 27-102   Cal.
   picta Hald. 47-58   Conn.-Ga.
   pilosa VanD. 27-101   Cal.
   sublineata Lec. 62-40   Pa. N.C.
   tenera Csy. 13-203   Cal.

Xylosteus Priv. 38-180 [12]
* ornatus Lec. 73-205   Cal. Ore.

Anthophylax Lec. 50-326 [3]
  (*Anthophilax* Lec. 50-236) [12] [3]
* hoffmanni Beut. 03-518   N.C.
   malachiticus Hald. 45-64   N.S.-N.C.-
   cyanea Hald. 47-151 [12]   [L.Sup.
   viridipennis Csy. 18-246 [12]
   viridis Lec. 50-326 [3] [12]
   attenuatus Hald. 47-59 N.B.-Va.-L.Sup.
   quadrimaculatus C. & K. 22-147   Chio
   mirificus Bland 65-382   B.C.-C.A.-Colo.
   venustus Bland 65-382 [12]
   costaricensis Bates 85-277 [12]
   tenebrosus Lec. 73-208   Ore. Cal.
   nigrolineatus VanD. 17-36 [12]
   subvittatus Csy. 91-37   Colo.

Pachyta Dej. 21-112 [12] [13]
* armata Lec. 73-207   B.C.-Or.-Ida.
   lamed Linn. 58-391   Alas.-Cal.-Pa.-
   liturata Kby. 37-178 [12]   [Que. Eur.
   nitens Lec. 50-235 [12]
   conflagrata Mots. 60-147 [12]

Piodes Lec. 50-318 [12]
* coriacea Lec. 50-318   Ore.

Evodinus Lec. 50-325 [12]
* monticola Rand. 38-27 N.S.-N.C.-Wis.
   carolinensis Csy. 24-275 [12]
   vancouveri Csy. 13-216   Alas.-Cal.

Gaurotes Lec. 50-324 [12]
* cressoni Bland 64-69   B.C.-Cal.-Colo.
   lecontei Csy. 13-219 [12]
   cyanipennis Say 23-423   N.B.-Ind.
   servillei Serv. 35-214 [12]
   ione Newm. 40-423 [12]
   laportei Guer. 44-253 [12]
   leonardii Hald. 47-60 [12]
   abdominalis Bland 62-270   Ont.-Va.

Ophistomis Thoms. 57-319 [12]
*   (*Cyphonotida* Csy. 13-260) [12]
   lævicollis Bates 80-39 Ariz. Mex. Guat.
   ventralis Horn 94-401 [12]

Bellamira Lec. 73-328 [12]
* scalaris Say 27-278   N.W.T.-Que.-Fla.
   coarctatus Hald. 47-59 [12]

Neobellamira Sw. & Hopp. 28-15 [12] [14]
* delicata Lec. 74-97   Cal.
   sequoiæ Hopp. 34-115   Cal.
   ?antennata Schffr. 08-342 (Strangalia) [12]

Strangalina Auriv. 12-240 [12] [14] [15]
*   (*Ophistomis* of Csy.) [12]
    (*Strangalia* of Lec.) [12] [14]
   virilis Lec. 73-212   Tex.
   acuminata Cliv. 95-20   Ont.-Md.-O.
   emaciata Newm. 41-68 [12]
   unicolor Hald. 47-62 [12]
   famelica Newm. 41-68   N.Y.-Ga.-Mo.-
   confluenta Hald. 47-61 [12]   [Minn.
   flaviceps Hald. 47-61 [12]
   obsoleta Hald. 47-61 [12]
   solitaria Hald. 47-61 [12]
   carolina Csy. 13-277 [13]
   ochreipennis Csy. 13-278 [12]
   bicolor Swed. 87-197   N.Y.-Ga.-Mich.
   simulans Csy. 13-278 [13]
   sexnotata Hald. 47-61   ?Mass. Fla.-
   texana Csy. 13-276 [12]   [N.M.-Colo.
   evanescens Csy. 13-276 [12]
   montana Csy. 91-40 [12]
   strigosa Newm. 41-69   Fla.
   luteicornis Fahr. 75-197   Mass.-Fla.-
                  [Tex.-Minn.

Euryptera Serv. 25-688 [12]
* ignita Schffr. 08-341   Ariz.
   huachucæ Schffr. 05-164   Ariz. Tex.
   lateralis Oliv. 95-22   Mass.-Fla.-Miss.
   distans Germ. 24-524 [12]   [Mex.
   cincta Hald. 47-63 [12]
   obsoleta Hald. 47-63 [12]
   lateralis Hald. 47-63 [12]
   flavatra Blatch. 14-92 [12]
   subintegra Csy. 24-285 [12]
   cruentata Martin 30-70   Ariz.

* * *

---

[9] Linsley—38.
[12] Revision of part of tribe. Hopping—37.
[13] Linsley — 38 changes the authority of these genera to Zetterstedt 28-376 (74); however, they are amply validated by Dejean in 1821 (p. 112) and must be credited to him.

[14] Hopping—34.
[15] This synonymy is given by Linsley—38 thus: Strangalia Serv. 35-220 (Strangalina Auriv. 12-228, Ophistomis Csy. 13-248).

Argaleus Lec. 50-319
  nitens Lec.,[16] for 50-235 read 50-319 [3]

Acmæops Lec., for 50-235 read 50-321 [3]
Leptura
  tibialis Lec., for 50-339 read 50-329 [3]
  splendens Knull 35-191   Ariz.

Anoplodera
  insignis Fall. 07-251 [6] Cal., Is. off L.Cal.
  vittata Cliv. 92-523   N.A.
    v.saratogensis Rau 35-63   N.Y.

Typocerus
  standishi Knull 38-141   Tex. Ckla.

              PSEBIINI [17]

Leptidiella Strand 36-169
    (Leptidea Muls. 39-105)
  brevipennis Muls. 39-105 [18]   Eur. Cal.

            * * *

Callimellum Strand 28-2
    (Callimus || Muls. 46-App.)[19]
  cyanipenne Lec. 73-192
    variipes Csy. 12-311 [3]
    dehiscens Csy. 12-312 [3]
  ruficolle Lec. 73-192
    longicolle Csy. 12-310 [3]
    s.opacipennis Csy. 12-311 [3]   Cal.

Pœcilobrium
  chalybeum Lec. 73-189   B.C.-Cal. Ida.
    rugosipenne Linell 96-395 [3]
    minutum Csy. 24-261 [3]
    gibsoni Hopp. 31-234 [3]

Semanotus
  bifasciatum Fahr. 87-152 [20]   N.A.
      ANACOMIS Csy. 12-271 [21]
  ligneus Fahr. 87-153 [21]   N.A.
    nicolas White 55-321
  litigiosa Csy. 92-25 [21]   Cal.
    v.terminata Csy. 12-274   Me.
      terminalis VanD. 37-112

Phymatodes
  blandus Lec. 50-79   Cal.
    s.picipes Linsley 34-165   Cal.
  propinquus Linsley 34-181   Cal.
  concolor Linsley 34-181   Cal.
  elongatus Hopp. 35-8   B.C.

[6] Linsley—34.
[16] Transferred from Pachyta, Linsley—38.
[17] To follow Necydalini, Linsley—33.
[18] Linsley—33.
[19] This synonymy is cited by Mequignon—37 but
fails to consider the prior name Pilema Lec. 73-
192 (generally considered to be a synonym) and
the still earlier name Lampropterus Muls. 63-214
(used as subgenus); nomenclaturally Lampropterus must be accepted as the genus name on the
basis of these facts.
[20] Transferred from Callidium, Linsley in litt.
[21] Van Dyke—37.

rainieri VanD. 37-113   Wash.
lecontei Linsley 38-109   Cal.
    obscurum || Lec. 59-79 [3]
    ?grandis Csy. 12-277 [3]   Cal.
  vulneratus Lec. 57-60   Cal. B.C.
    nigrescens Hardy & Pr. 27-190 [22]

Xylotrechus
  undulatus Say 24-291   L.Sup. Id.
    fuscus Kby. 37-176 [21]
    lunulatus Kby. 37-175 [21]
  frosti VanD. 37-114
    fuscus of authors [21]
  insignis Lec. 73-199   Cal.
    v.nunenmacheri VanD. 30-43 [4]   Ore.

Neoclytus
  confusus VanD. 37-115
    kirbyi of authors [21]
    longulus of authors [21]
  nubilus Linsley 33-93   Cal.
  muricatulus Kby. 37-177   Can. N.St.
    kirbyi Auriv. 12-392 [21]
    longipes || Kby. 37-176 [21]
  acuminatus Fabr. 75-194.   E.N.A.
    s.hesperus Linsley 35-163   Colo.
  resplendens Linsley 35-163   Cal.

Triodoclytus Csy. 13-387
    (Synclytus Lucas 20-480)[23]

Euderces
  balli Knull 35-192   Ariz.

Rhopalophora
  bicolorella Knull 34-336   Ariz. Tex.

Stenosphenus
  aridus Linsley 35-166   Utah
  arizonicus Linsley 35-165   Ariz.
  basicornis Linsley 34-60   L.Cal.

Atimia
  helenæ Linsley 34-25   Cal.

Elytroleptus
  floridanus Lec. 62-42   Mass. N.Y. Fla.
    s.immaculipennis Knull 35-99   Tex.

Tragidion
  opacum Knull 37-306   Tex.

Moneilema Say 24-403
    (Monoplesa Mots. 75-144)[23]
  annulata Say 24-404   Kan.-Mont.-Ar.
    armigera Mots. 75-146 [23]   Cal.
    scabra Mots. 75-146 [23]   Cal.

Monochamus
  fulvomaculatus Linsley 33-118   Cal.

Dorcaschema
  octovittata Knull 37-307   Tex.

[4] Linsley—35.
[22] Hopping—35.
[23] Linsley in litt.

Alphomorphus Linsley 35-100 [24]
  vandykei Linsley 31-79 (Pogonocherus)[4]
              [Tex.

Leptostylus
  falli Linsley 34-182   Ariz.
  monki Knull 36-106   Tex.

Leiopus
  imitans Knull 36-107   Tex.

      POGONOCHERINI [25]

Zaplous Lec. 78-418 [25]
* annulatus Chev. 62-250   Fla. Cuba
  hubbardi Lec. 78-415 [25]

Lypsimena Lec. 52-155 [25]
*   (*Allœoscelis* Bates 85-358)[25]
  fuscata Lec. 52-155   N.Y.-Fla. Cuba,
    leptis Bates 85-358 [25]   [Mex.-S.A.
  californica Horn 85-194   Cal.

Callipogonius Linsley 35-79 [25]
* cornutus Linsley 30-86   Tex.

Poliænus Bates 80-120 [25]
  (*Pogonocherus* of auth., in part)[25]
  californicus Schffr. 08-347   **Cal.**
  pilatei VanD. 20-46 [25]
  concolor Schffr. 09-102   Cal.
  albidus Linsley 34-184   Cal.
  oregonus Lec. 61-354 Pac.Cst.-Rky.Mts.
  obscurus Fall 10-5   Ariz.
    s.ponderosæ Linsley 35-85   Cal.
  schaefferi Linsley 33-184   Cal.
    vandykei Schffr. 32-153 [4][25]
    californicus of authors (in part)[4][25]
  negundo Schffr. 08-164   Ariz.
  volitans Lec. 73-232   L.Cal. Guat.
  hirsutus Bates 80-120 [25]

Ecyrus Lec. 52-160 [25]
* penicillatus Bates 80-137   Tex. Mex.
  fasciatus Hamilt. 96-137 [25]
  texanus Schffr. 08-347   Tex.
  dasycerus Sav 27-270   E.N.A. Can.
  obscura Hald. 47-50 [25]
  exiguus Lec. 52-161 [25]
  s.floridanus Linsley 35-93   Fla.

Lophopogonius Linsley 35-94 [25]
* crinitus Lec. 73-267   Pac.Cst.

Pogonocherus Zett. 28-364 [25]
  (*Pityphilus* Lac. 72-653)[25]
    POGONOCHERUS (s.str.)[25]
  penicillatus Lec. 50-234 Alas. Rky.Mts.
  alaskanus Schffr. 08-385 [25] [E.N.A.

EUPOGONOCHERUS Linsley
            [35-97
  propinquus Fall 10-6 Pac.Cst. Rky.Mts.
  arizonicus Schffr. 08-346   Ariz.
  medianus Linsley 35-98   Ariz.
  pictus Fall 10-6   Rky.Mts. N.Pac.Cst.
  simplex Hamilt. 96-135 [25]
  emarginatus Csy. 13-347 [25]
  fastigiatus Csy. 13-348 [25]
  mixtus Hald. 47-50   N.N.A.
  simplex Lec. 73-237 [25]
  parvulus Lec. 52-160   N.N.A.
  salicicola Csy. 13-347 [25]

      • • •

Lochmæocles
  tesselatus Thoms. 67-90 [26] Tex. S.A.

Cylindrataxia Linsley 34-183 [27]
  salicicola Linsley 34-184   Tex.

Adetus Lec. 52-161 [28]
*   (*Sicyobius* Horn 80-137)[6]
  vanduzeei Linsley 34-62   L.Cal.
  brousi Horn 80-137   Tex.

CHRYSOMELIDÆ

Aulacoscelis
  ventralis Schffr. 33-297   Ariz.
  femorata of Schffr.[1]

Donacia
  subtilis
    s.magistrigata Mead 38-113 Cal.
  idola Hatch 38-110   Wash.
  distincta
    s.occidentalis Mead 38-114   Cal.
  germari Mannh. 43-306   Ore.
  flavipennis Mannh. 43-306 [2]
  emarginata Kby. 37-224   N.A.
  pacifica Schffr. 25-135 [3]

Crioceris
  duodecimpuncta
    v.dodecastigma Suffr. 41-40 [1] Eur.
                    [N.A.

Lema Fabr. 98-90 [3]
* peninsulæ Cr. 73-24   L.Cal.
  margineimpressa Schffr. 33-299 Ariz.
  concolor Lec. 85-24   N.Mex.
  texana Cr. 73-24   Tex.
  arizonæ Schffr. 19-320   Ariz.
  sayi Cr. 73-25   S.E.U.S.
  cornuta Fahr. 01-475   S.E.U.S.
  simulans Schffr. 33-300   Kans.
  coloradensis Linell 98-475   Colo.
  gaspensis Brown 38-35   Que.
  palustris Blatch. 13-22   E.U.S.
  v.floridana Schffr. 33-300 [3]   Fla.

---

[4] Linsley—35.
[6] Linsley—34.
[24] Belongs in tribe Acanthoderini near Alphus, Linsley—35.
[25] Revision of tribe. Linsley—35.
[26] Knull—37.
[27] Belongs in Ataxiini near Aporataxia, Linsley —34.
[28] Belongs in Adetini (to follow Ataxiini), Linsley—34.
[1] Schæffer—33.
[2] Mead—38.
[3] Revision of genus, Schæffer—33.

brunnicollis Lac. 45-391 Fla.
maculicollis Lac. 45-392 S.E.U.S.
collaris Say 24-430 Kan.-Ind. Fla.
longipennis Linell 98-474 Colo. Neb. Ill.
conjuncta Lac. 45-408 Fla.
 v.circumvittata Clark 66-41 [3] Fla.
solani Fahr. 98-93 S.E.U.S.
confusa Chev. 35-166 Fla.
 v.trabeata Lac. 45-409 [3] Fla. Ariz.
 v.omogera Horn 94-405 [3] L.Cal.
balteata Lec. 85-24 Ariz.
 v.equestris Lac. 45-403 [3] Ariz.
melanocephala Say 27-294 N.W.U.S.
opulenta G. & H. 74-3258 Tex.
flavida Horn 94-405 L.Cal.
nigrovittata Guer. 29-262 N.M. Ariz.
trilineata Oliv. 08-739 E.U.S.
 v.medionata Schffr. 33-303[1] [3] Fla.-N.C.
 v.trivittata Say 24-429 [1] [3] [4] W.St. N.Y.-
 immaculicollis Chev. 35-112[1] [Ala.
 trivirgata Lec. 59-22 [1]
 nigrovittata of Schffr.[1]
 v.californica Schffr. 33-301 [31] Cal.
 nigrovittata Guer. 29-262 N.M.-
 notativentris Schffr. 19-322[1] [C.A.
æmula Horn 94-406 L.Cal.
jacobina Linell 98-474 Tex.
sexpunctata Cliv. 08-738 S.E.U.S.
 v.albini Lac. 45-492 [3] S.E.U.S.
 v.ephippium Lac. 45-483 [3] S.E.U.S.

(*Antipus* DeG., *not North American*)[1]

Anomœa Lac. 48-130 [5] [6]
* mutabilis Lac. 48-137 Tex. Mex.
 ruficauda Fœrsb. 21-261 [6]
 nitidicollis Schffr. 19-322 Tex.
 crassicornis Schffr. 33-313 Fla.
 laticlavia Forst. 71-27 Atl.St.-S.D.-S.A.
 v.floridana Schffr. 33-315 [6] Fla.
 v.kansana Schffr. 33-314 [6] Kan.
 angustata Schffr. 33-316 Fla.
 högei Jacoby 88-66 Tex. Mex.

Gynandrophthalma Lac. 48-256 [6]
* militaris Lec. 58-83 Tex.
 arizonica Schffr. 19-323 Ariz.

Coscinoptera
 dominicana Fahr. 01-34 N.Eng.-Fla.-
 [Mex.
 v.franciscana Lec. 59-22 [1] Tex. Ariz.
 dorsalis Lec. 74-25 [1] [Colo. Kan.

Babia Lac. 48-424 [6]
* quadriguttata Cliv. 91-37 E.U.S.-Neb.-
 [Tex.
 v.pulla Lac. 48-429 [6] Ariz. N.M.
 v.tenuis Schffr. 33-320 [6] Wyo.
 tetraspilota Lec. 58-83 Ariz. N.M.
 v.texana Schffr. 33-320 Tex.
 humeralis Fabr. 01-37 L.Cal. Mex.
 oregona Schffr. 33-321 Ore.

[4] Also cited as a valid species, Schæffer—33.
[5] Cited as valid genus, Schæffer—33. (Titubœa
Lac. is listed as a possible synonym).
[6] Revision of genus. Schæffer—33.

Fulcidacinæ [1]

(Chlamydinæ)[1]

Arthrochlamys Ihering 05-642
 (Chlamys) || Knoch 01-122)[1]

Monachulus
 scaphidioides Suffr. 51-215 [7] Mex. Ariz.
 opacicollis Schffr. 33-321 Ariz.

Cryptocephalus
 notatus
 v.sellatus Schffr. 33-322 Tex.
 binominis
 v.rufibasis Schffr. 33-322 Fla.
 multisignatus Schffr. 33-322 Ariz.
 trizonatus Suffr., for Ariz. read Tex.[1]
 egregius Schffr. 34-459 Tex. Ga.
 snowi Schffr. 34-461 Ariz.
 cowaniæ Schffr. 34-462 Ariz.
 duryi Schffr. 06-230 [1] Ariz. Tex.
 cupressi Schffr. 33-324 La.
 texanus Schffr. 33-323 Tex.
 amatus Hald. 49-253
 apicidens Fall 32-22 [7]
 spurcus Lec. 58-84 Cal.
 s.vandykei White 37-112 Cal.
 cerinus White 37-111 Cal.
 s.nevadensis White 37-113 Nev.
 incertus Hald. 49-250 Ga. Conn.
 calidus Suffr. 51-241 [8]
 bispinus Suffr. 58-347 Fla.
 albicans Hald. 49-252 [8]
 pumilus Hald. 49-249 Ga. Fla.
 pseudolus Suffr. 58-373 [5]
 simulans Schffr. 06-231 Ariz.
 v.conjungens Schffr. 34-460 Tex.
 v.eluticollis Schffr. 34-460 Ariz.
 luteolus Newm. 40-250
 defectus Lec. 80-201 [8]
 sanfordi Blatch. 13-23 [8]
 sanfordensis Clav. 13-182 [5]

Diachus
 luscus Suffr. 58-377 [9] Ga.

Bassareus
 formosus Melsh. 47-173
 v.egenus Suffr. 51-311 [5] Pa.
 v.confluentinus Schffr. 34-464 [8] Mass.-
 [Pa.
 mammifer Newm. 40-250
 speciosus Schffr. 34-463 (in error)[8]
 pretiosus Melsh. 47-174
 lituratus Fahr. 01-50
 v.geminatus Hald. 49-253 [9]
 vittatus Suffr. 51-296 [8]

Nodonota
 basalis Jacoby 90-197 Mex. Ariz.
 arizonica Schffr. 06-238 [8]

[7] Fall—34.
[8] Schæffer—34.
[9] Transferred from Cryptocephalus. Schæffer—34.

[8] Schæffer—34.
[10] Described from paratypes of flavocostata, Schffr.; this whole group badly needs revision; brunnea Fahr. must be changed. Barber—37.
[11] Revision of genus. Schæffer—34.
[12] Revision of genus. Krauss—37.

[13] Revision of genus, Van Dyke—38.
[14] Leng Catalog.
[15] Revision of genus, Blake—36.
[16] Transferred from Monoxia, Blake—36.

Disonycha Chev. 37-414 [17]
* pennsylvanica Ill. 07-146   Mass.-Fla.-
  sexlineata Cliv. 08-642 [17]   [Tex.-Ill.
  pensylvania Strum. 43-283 [17]
  parva Blatch. 22-16 [17]
conjugata Fabr. 01-495 N.C.-Fla. Cuba
  costipennis Duval 57-129 [17]
  floridana Jacoby 01-146 [17]
procera Csy. 84-182 B.C.-H.B.T.-Ga.-Ut.
  ?vicina Kby. 37-217 [?]
  ?pallipes Cr. 73-64 [17]
  pennsylvanica of Horn 89-202
                        [(part)[17]
  nigriventris Schffr. 31-282 [17]
uniguttata Say 24-88   Mass.-Fla.-La.-
  ?vicina Kby. 37-217 [17]   [Man.
  ?pallipes Cr. 73-64 [17]
  pennsylvanica of Horn 89-202
                        [(part)[17]
limbicollis Lec. 57-67   Cal. Nev.
alternata Ill. 07-144 N.W.T.-N.S.-S.C.-
  ?quinquevittata Say 24-88 [17] [Cal.
  punctigera || Schffr. 31-279 [17]
latiovittata Hatch 32-108 B.C.-Cal. Wyo.
  puncticollis || Lec. 57-67 [?]   1
  quinquevittata Horn 89-203 (part)[17]
schaefferi Blake 33-24   Can. O.
pluriligata Lec. 58-27   Ill.-La -Colo.
  ?quinquevittata Say 24-88[17]
  v.pura Lec. 58-86 [17] Cal.-N.M. C.A.
  capitata Jacoby 84-316 [17]
  quinquevittata of Jacoby 91-276[?]
punctigera Lec. 59-24   Ill.-N M.-Alb.
  ?quinquevittata Say 24-88[17]
  neglecta Schffr. 31-283 [17]
  punctipennis Schffr. 31-284 [17]
arizonæ Csy. 84-52 Me.-S.C.-Ariz.-Man.
  glabrata Jacoby 84-311
  davisi Schffr. 24-141
  v.borealis Blake 33-31   Ont. Mich.
tenuicornis Horn 89-208   Ariz. N.M.
caroliniana Fabr. 75-122   Mass.-Fla.-
  ?s-vittata Fabr. 92-47 [17]   [Tex.-Ill.
  ?vittata Fabr. 01-491 [17]
  quinquevittata G. & H. 76-3497 [17]
  pulchra Csy. 84-51 [17]
  alternata || Horn 89-315 [17]
figurata Jacoby 84-314 Nev. Ariz. C.A.
fumata Lec. 58-86   Tex. Mo. ?N.Y.
  crenicollis of Horn 89-204 (part)[17]
  alternata Jacoby 84-311 (part)[17]
  horni Jacoby 84-311 (part)[17]
  v.quinquerutata Schffr. 19-336 [17] Cal.-
  carolina Brisley 25-175[17][N.M. Ut.
  v.lodingi Schffr. 19-337 [17]   Ala.
latifrons Schffr. 19-336 Cal.-N.M.-Mont.
  v.laticollis Schffr. 31-284 [17] N.S.-N.J.-
                        [Minn.
  quinquevittata of Whiteh. 18-38 [17]
discoidea Fabr. 92-25   Md.-Ga.-Tex.-
  nigridorsis G. & H. 76-3497 [17] [Kan.
  v.abbreviata Melsh. 47-163 [17] Pa.-Va.-
                        [Ill.

[17] Revision of genus, Blake—36.

leptolineata Blatch. 17-143   Fla.
  v.texana Schffr. 19-339 [17] Va.-Ariz.
                        [Kan.
antennata Jacoby 84-35   Fla. Mex.
  albida Blatch. 24-169 [17]
alabamæ Schffr. 19-337   Ala. Tex.
admirabilis Blatch. 24-90   Mass.-Va.-
                        [Tex.-Ind.
glabrata Fabr. 81-156 N.Y.-Fla.-Ill.-S.A.
  tomentosa of Fabr. 75-122 [17] [W.I.
  vittata Cliv. 89-105 [17]
  alternata of Latr. 33-39 [17]
  horticola Dej. 37-414 [17]
  albicollis Sturm. 43-283 [17]
maritima Mannh. 43-311   Cal. Nev.
collata Fahr. 01-463 Me.-Fla.-Ill.-C.A.
  collaris Cr. 73-64 [17]
  mellicollis of Horn 89-211 [17]
semicarbonata Lec. 59-25   N.M. Colo.
  mellicollis Horn 89-211 (part)[17]
xanthomelas Dalman 23-79   N.W.T.-
  collaris of Ill. 07-126 [17] [Mass.-Va.-
  xanthomelæna G.&H. 76-3497 [17][La.
  merdivora Melsh. 53-122 [17]
  v.cervicalis Lec. 59-25 [17]   Kans. Ga.
  v.atrella Blake 33-57   Mass. Va. Ala.
triangularis Say 24-84   B.C.-Mass.-Va.-
  puncticollis Kby. 37-218 [17]   [Tex.
  v.montanensis Blake 33-59   Mont.
politula Horn 89-211   Kan.-N.M.-Guat.
varicornis Horn 89-210 Tex. Cal. L.Cal.
funerea Rand. 38-47 Mass. Ct. Ga. Ala.
brevicornis Schffr 31-281   Colo.
stenosticha Schffr. 31-285   Tex.
quinquevittata Fabr. 77-118 [18]   Carol.
  s-vittata Fahr. 92-47
mellicollis Say 31-10 [18]   La. Mo.
vicina Kby. 37-217 [18]   Can.
pallipes Cr. 73-64 [18]
quinquevittata Say 24-85 [18]   Mo.

Altica [19]
bimarginata Say 24-85 [20]
plicipennis Mannh. 43-310   Alas.-Cal.-
                        [Tex.-Man.
subplicata Lec. 59-25   Can.-La.-N.M.-
                        [Minn.
prasina Lec. 57-67   B.C.-Cal.-Ut.
ambiens Lec. 59-25 B.C.-Cal.-N.M.-Wyo.
  v.alni Harris 69-268 [19]   Me-Minn.
  v.latiplicata Blake 36-21   Ore. Cal.
                        [Ariz. Nev.
guatemalensis Jacoby 84-297 Ariz. C.A.
napensis Blake 36-23   Cal.
caurina Blake 36-24   Alb. Wash.
  * * *
woodsi Isely 20-11 [21]   Iowa
nancyæ Stirrett 33-208   Iowa
populi Brown 38-37   Ont. Mann.

[18] Unrecognized species, Blake—33.
[19] Revision of part of genus, Blake—36.
[20] Unrecognized species, Blake—36.
[21] Dietrich in litt.

Orestioides Hatch 35-276 [22]
* robusta Lec. 74-274 [23] N.H. Que. Wash.

Systena
elongata Fabr. 98-99   N.Y.-Fla.-Cal.
  subœnea Lec. 57-68 [24]   [Man.
dimorpha Blake 33-181 Cal.-N.M.-Kan.-
                                [Mont.

- - -

tæniata Say 24-294 [24]
blanda Melsh. 47-164 [24] N.Y.-Ga.-N.M.-
                                [Ida.
  s.ligata Lec. 57-68 [24]   Cal.
  ochracea Lec. 58-87 [24]
mitis Lec. 58-87 [24]   Cal.-Tex.-Colo.
bitæniata Lec. 59-26 [24] Cal.-N.M.-Alb.-
                                [Dak.
(pallidula Boh., not North American) [24]
lævis Blake 35-100   Cal.-Colo. Ariz.
californica Blake 35-101 [24]   Cal.
carri Blake 35-102 [24]   Alb.

* * *

gracilenta Blake 33-180   Tex. Mex.

Psylliodes
credens Fall 33-233   Cal.
verisimilis Fall 33-232   N.Mex.

Anisostena
arizonica Schffr. 33-103   Ariz.
texana Schffr. 33-103   Tex.
kansana Schffr. 33-104   Kans.

Anoplitis
ancoroides Schffr. 33-105   N.J.

Chalepus Thunb. 05-282
* bicolor Cliv. 92-96 [25]   Tex. E.St.
waishi Cr. 73-81 [25] Ill.-Colo.-Ark.-Ariz.

Xenocephalus Weise 10-136 [26]
        HEMICHALEPUS Spæth 37-147 [26]
arizonicus Uhm. 38-425   Ariz. Mex.
  crotchi Weise 10-144 [26]

Brachycoryna
lateralis Schffr. 33-105   Colo.

Stenopodius Horn 83-301 [27]
* flavidus Horn 83-301 [27]   Cal.
submaculatus Blais. 39-433   Cal.
  v.laticollis Blais. 39-434   Cal.
  v.pallidulus Blais. 39-435   Cal.
inyoensis Blais. 39-435   Cal.
  v.pallidus Blais. 39-437   Cal.
martini Blais. 39-437   Tex.
texanus Schffr. 33-106 [27]   Tex.
vanduzeei Blais. 39-440   Cal.
insularis Blais. 39-442   Gulf of Cal. Is.

[22] Belongs in Halticini near Crepidodera and Orestia. Hatch—35.
[23] Transferred from Crepidodera, Hatch—35.
[24] Revision of part of genus, Blake—35.
[25] Uhmann—36.
[26] Uhmann—38.
[27] Revision of genus, Blaisdell—39.

Pentispa
suturalis Baly 85-51   Mex. Ariz.
  (v.vittula Weise, not No. American)[1]

Microrhopala
arizonica Schffr. 06-253 [28]

Chelymorpha
phytophagica Cr. 73-77   Ariz.
  v.luteata Schffr. 33-108   Ariz.

Cassida
(nebulosa Linn., not North American)[1]
flaveola Thunb. 94-103 [1] Eur. Pa. Md.
                                [N.Y.

Chirida
barberi Spæth 36-140   Fla.

Psalidonota Boh. 55-81
* texana Schffr. 33-108   Tex.
leprosa of Horn [1]

Erepsocassis Spæth 36-260
* rubella Boh. 62-449   Ala. Ga.
  marginepunctata Schffr. 25-236 [29]
  marginepuncta Spæth 36-260

Strongylaspis Spæth 36-216 [30]
* bisignata Boh. 55-119 [29] Kans. O. Tex.

BRUCHIDÆ

(Mylabridæ, Acanthoscelidæ)

Bruchus
tenuis Bott. 35-127   Fla.-Tex.-Mich.
brachialis Fahr. 39-79 [31]   O.-N.J.-Ga.

BRENTIDÆ

Arrhenodes
minutus Drury 70-95 [1]   E.N.A.

PLATYSTOMIDÆ

(Anthribidæ)[2]

Goniocloeous Jord. 04-260
*
    (Tropideres of authors)[3]
bimaculatus Cliv. 95-14 [3]   E.St.
rectus Lec. 76-395   E.St.
Brachytarsoides Pierce 30-29
irregularis Tanner 34-285   Utah

[1] Schæffer—33.
[28] Probably valid species, Schæffer—33.
[29] Spæth—36.
[30] To follow Metriona, Spæth—36.
[31] Bridwell & Bottimer—33; Bottimer—36 and 37.
[1] Transferred from Eupsalis (or Platysystrophus) by Buchanan—39.
[2] This name used by Zimmerman—36 and Wolfrum—38.
[3] Transferred from Tropideres by Wolfrum—38.

CURCULIONIDÆ

(Auletes Schönh., not North American)[4]

Auletobius Desbr. 68-396 [4][5]
  (Metopon Waterh. 42-lxii)[4]
  (Involvulus Schrank 98-476)[4]
  (Auletes of authors)[4]
  AULETOBIUS (s.str.)[4]
  congruus Walk. 66-331 [4][5] Ida-Neb.-Colo.
  subcoeruleus Lec. 76-4 [5]
    MESAULETES Voss 33-116 [5]
  nasalis Lec. 76-412 [4][5]   Cal.
  humeralis Boh. 59-117 [4][5]   Cal.
  rufipennis Pierce 09-327 [5]
  albovestita Bl. & Leng 16-54 [4][5]   N.A.
    ALLETINUS Desbr. 08-79 [4][5]
    (Nemonus Desbr. 08-13)[5]
  ater Lec. 76-4 [4]-   Mass.-Tex.-Ont.

* * *

?cassandræ Lec. 76-5 [6]   N.A.
?laticollis Csy. 88-233 [6]   Cal.
?blatchleyi Voss 35-240 [6]   Fla.
  minor Blatch. 22-98 [6]
?viridis Pierce 09-327 [6]   Cal.

Deporaus Leach 19-201
  * (Platyrhynchus Thunb. 15-123)[7]
    HYPODEPORANUS Voss
                          [23-388 [7]
  glastinus Lec. 57-52 [7] Colo. Ariz. Cal.
                          [Wash.

Rhynchites Schneid. 91-82 [7]
    HAPLORHYNCHITES [7]
  eximius Lec. 76-413 [7]   Colo. Ariz.
  æneus Boh. 29-22 [7]   Can. Fla. Ore.
    INVOLVULUS Schrank 98-475[7]
  hirtus Fabr. 01-421 [7]   Mass. Fla. Mich.
  consobrinus Voss 38-159 [7]   N.A.
    RHYNCHITES (s.str.)[7]
  velatus Lec. 80-216 [7]   Cal.

Corigetus Desbr. 73-662
  ?castaneus Rœlofs 73-168 [8]   Asia, N.Y.

Brachyrhininæ

(Otiorhynchinæ)

Trichalophus [9]
  seminudus VanD. 38-7   Colo.

Plinthodes [9]

Triglyphulus Ckll. 06-243 [9]
  * ater Lec. 76-117   Cal.
  nevadensis VanD. 38-8   Nev.

Acmægenius [9]

Sapotes
  longipilis VanD. 34-175   Ariz.

Eupagoderes
  huachucæ VanD. 34-180   Ariz.
  halli VanD. 34-181   Ariz.
  simulans VanD. 34-176   Tex.
  decipiens Lec. 53-445
    dunnianus Csy. 88-240 [10]
  ocellatus VanD. 34-177   Colo. Ut.
  setosus VanD. 34-179   Ariz.

Cimbocera Horn 76-55 [11]
  * cinerea VanD. 35-1 [11]   Colo.
  cazieri VanD. 36-73
  pauper Horn 76-56 [11] Dak. Wyo. Mont.
  conspersa Fall 07-261 [11]   Mont.-N.M.-
    sericea Pierce 13-379 [11][12]   [Ariz.
  robusta VanD. 35-2 [11]   Cal.

Paracimbocera VanD. 38-1
  * atra VanD. 38-2   Nev.

Miloderoides VanD. 36-74
  * maculatus VanD. 36-76   Ida.
  argenteus VanD. 35-4 [13]   Colo.
  Lepidopus VanD. 36-76
  * nevadicus VanD. 36-77   Nev.
  parvulus VanD. 36-78   Ida.

Dichoxenus
  setosus Blatch. 16-103 [14]   Ind. Mo.

Pseudorimus VanD. 34-182
  * granicollis VanD. 34-183   Ariz.
    gravicollis VanD. 34-185
  orbicollis VanD. 34-184   N.Mex.

Crocidema VanD. 34-185
  * californica VanD. 34-187   Cal.
  nigrior VanD. 34-188   Ariz.
  planifrons VanD. 34-189   Ariz.
  attenuata VanD. 34-190   Utah
  albovestita VanD. 34-190   Ariz.

Melanolemma VanD. 35-5
  montana VanD. 35-5   Colo.

Peritaxia Horn 76-46
  (Parataxia VanD. 36-79)
  uniformis VanD. 36-79   Ariz. Colo.
  brevipilis VanD. 35-6   Ariz.

---

[4] Voss—33.
[5] Revision of genus, Voss—34.
[6] Doubtful species, Voss—35.
[7] Voss—38.
[8] Davis—35.
[9] Transferred from Alophini, Wilcox & Davis—35.

[10] Van Dyke—34.
[11] Revision of genus, Van Dyke—35.
[12] Listed as variety by Van Dyke in 1934.
[13] Described in Miloderes, Van Dyke—35.
[14] Transferred from Anametis, Van Dyke—34.

Dyslobus Lec. 69-380 [15]

*   (*Amnesia* Horn 76-48)[16]
    (*Thricomigus* Horn 76-48)[16]
    (*Melamorphus* Horn 76-40)[10][16]

segnis Lec. 57-56	Cal. Ore.
lecontei Csy. 95-811	Cal-Wash.
simplex VanD. 33-37	Ore-Wash.
verrucifer Csy. 95-812	Ida-B.C.
bituberculatus Pierce 13-388 [15]	
denticulatus Pierce 13-388	Cal.
alepidotus Ting 37-79	Cal.
granicollis Lec. 69-380	B.C.-Cal.
sculptilis Csy. 88-250 [15]	
discors Csy. 95-814 [16]	
debilis Csy. 95-815 [15]	
s.tumidus Csy. 95-813[16]	Cal.
s.vestitus VanD. 33-39	Cal.
viridescens VanD. 33-33	Ore.
squamipunctatus Pierce 09-350	Cal.
wilcoxi VanD. 33-40	Ore.
decoratus Lec. 69-381	Cal-B.C.
ursinus Horn 76-51	Ore.
raucus Horn 76-51	Cal. Vanc.
ciliatus Pierce 13-385 [15]	
tanneri VanD. 33-42	Utah
wasatchensis Tanner 38-147	Utah
argillous VanD. 35-7	Utah
remotus VanD. 3^8-3	Ore.
blaisdelli VanD. 33-42	Cal.
bakeri VanD. 33-43	Cal.-Wash.
luteus Horn 76-48	Colo. Mont. Wash.
granulatus Csy. 88-248	Cal. Ore.
deciduus Horn 76-52	Cal.
sordidus Horn 76-52	Cal.
elongatus Horn 76-53 [16]	
alternatus Horn 76-52	Mont. Alb. [Wash. Cal.
tessellatus Csy. 88-249	Cal.
franciscanus VanD. 33-45	Cal.
nigrescens Pierce 13-384	Wash.
niger Horn 76-40	Nev.
dolorosus VanD. 33-46	Cal.

Adaleres

flandersi VanD. 35-8	Cal.

Panscopus Schönh. 42-266 [17]

    PANSCOPUS (s.str.)

alternatus Schffr. 08-214	N.C.Ga.W.Va.
erinaceus Say 31-12	N.H.-Va.-Wis.
carinatus Pierce 13-398 [17]	
impressus Pierce 13-395	S.E.St.
s.thoracicus Buch. 36-4	N.C.
alternatus Pierce 13-394 [17]	

    PARAPANSCOPUS Buch. 36-[5 [17]

maculosus Blatch. 16-105	N.Y.-Ky.-Ia.
ovatipennis Buch. 36-6	Ont.
s.verrucosus Buch. 36-7	Pa.

    PHYMATINUS Lec. 69-382 [17]

gemmatus Lec. 57-56	Cal.-Wash.

NOCHELES Lec. 74-453 [17]

    (*Panscopideus* Pierce 13-[394)[17]

torpidus Lec. 57-55	Wash. Ore.
squamosus Pierce 13-394	Ore.
v.dentipes Pierce 13-395 [17]	Wash.
michelbacheri Ting 38-121	Cal.

    DOLICHONOTUS Buch. 36-9 [17]

convergens Buch. 36-10	Ore.
oregonensis Buch. 36-11	Ore.

    PSEUDOPANSCOPUS Buch. [27-33 [17]

costatus Buch. 27-33	B.C. Wash.

    NEOPANSCOPUS Pierce 13-[397 [17]

æqualis Horn 76-55	Kan.-Alb.-B.C.-Cal.
cinereus Horn 76-55 [17]	
vestitus Csy. 88-251 [17]	
squamifrons Pierce 13-397	Cal.
wickhami Buch. 36-13	Cal.

    NOMIDUS Csy. 95-818 [17]

johnsoni VanD. 35-9	Wash.
bufo Buch. 27-31	Cal.
abruptus Csy. 95-819	Cal. Wash.
rugicollis Buch. 27-31	Wash. Ore.
schwarzi Buch. 27-30	Ut. Ida.
longus Buch. 36-16	Wash.
pallidus Buch. 27-31	B.C. Wash. Mont.
tricarinatus Buch. 27-32	Ore.
bakeri Buch. 36-17	Wash.
ovalis Pierce 13-396	Alb.

• • •

coloradensis VanD. 36-80	Colo.

Tanymecus

texanus VanD. 35-86	Tex.

(handwritten annotation)

Hormorus [18]

Agasphærops Horn 76-24 [18]

* nigra Horn 76-25	Cal.
sulcirostris Pierce [19]	

Lupinocolus VanD. 36-81 [18]

♦ blaisdelli VanD. 36-82	Cal. Nev.

• • •

Mitostylus

elongatus VanD. 36-83	Tex.

Epicærus Schönh. 34-323

* (*Melbonus* Csy. 95-820)[19]	
scapalis Csy. 95-821 [19]	Ariz.
denticulatus Pierce 09-350 [19]	
uniformis Tanner 34-287	Utah

(handwritten annotation)

---

[10] Van Dyke—34.
[15] Revision of genus. Van Dyke—33.
[16] Used also as subgenera. Van Dyke—33.
[17] Revision of genus. Buchanan—36.
[18] Revision of new tribe in Brachyrhininæ. Van Dyke—36.
[19] Transferred from Panscopus (Phymatinus). Buchanan—36.

Mimetes Schönh. **47-23** [11]
* *(Amotus* Csy. 88-237)[10] [11]
  (Stamoderes Csy. 88-237)[10] [11]
setulosus Schönh. 63-40   Cal.
  gracilior Csy. 88-245 [11]
  longisternus Csy. 88-244 [11]
uniformis Csy. 88-237   Cal.
longipennis Pierce 09-348   Cal.
seniculus Horn 76-45   Cal.
lanei VanD. 35-86   Wash. Ore.
—Iasalanus Tanner 34-287   Utah

Pantomorus Schönh. 39-942 [20]
  GRAPHOGNATHUS Buch. 39-
               [11 [20]
leucoloma Boh. 40-62 [21] S.A. Australia,
                  [Fla.-La.
peregrinus Buch. 39-14   Miss.
  ATRICHONOTUS Buch. 39-15[20]
tæniatulus Berg 81-61   Fla.-Tex. S.A.
  texanus Pierce 11-49 (Artipus)[20]
  ASYNONYCHUS Cr. 67-388 [20]
    (Aramigus Horn 76-93)[20]
    (Aomopactus Jekel in Horn
              [76-94)[20]
godmani Cr. 67-389   N.A. S.A. C.A.
  fulleri Horn 76-94 [20] [Africa, Crient,
  olindæ Perkins 00-130 [20]   [Eur.
  ovulum Jekel, Hust. 22-100 [20]
  ?*subvittatus* Fairm. & Germ. 61-7[20]
tessellatus Say 24-318   Ill.-Ckla. Kan.
  sublineatus Dej. 37-289 [20]
  ?durius Boh. 40-27 [20]
  -*?candida* Horn 07-248 [20]
pallidus Horn 76-94   Ill.-Tex.-Colo.
  ?durius Boh. 40-27 [20]
  ?candida Horn 07-248 [20]
  PHACEPHOLIS Horn 76-95 [20]
elegans Horn 76-95
  v.elegans (s.str.)   Neb.-Tex.-Cal.
  metallicus Pierce 13-417 [20]
  v.viridis Pierce 09-361 [20]   Tex. Mex.
  v.pallidulus VanEmd. 36-28 [20]   Tex.
  pallidus || Pierce 10-363 [20] [22]
  v.eximius Buch. 39-32 [20]   Tex.
texanellus Buch. 39-36   Tex.
  texanus || Pierce 13-417 [20]
candidus Horn 76-97   Kan.-S.D.-Mont.-
  nebraskensis Pierce 13-416[20] [Colo.
planitiatus Buch.39-36 Neb.-N.M.-Mont.
obscurus Horn 76-96   Kan.-Tex.

Mesagroicus
elongellus VanEmd. 36-30   Ore.
  elongatus Buch. 29-10 [22]
  v.nevadicus Buch. 29-11 [22]   Nev.
  v.incertus Buch. 29-11 [22]   Wash.

Cryptolepidus VanD. 36-191 [23]
  (Lepidopus VanD. 36-76)

Trigonoscuta [24]
imbricata VanD. 36-83   Cal.

Thysanocorynus VanD. 38-5 [24]
aridus VanD. 38-6   Cal.

Stereogaster VanD. 36-84 [25]
globosa VanD. 36-85   Cal.

Otiorrhynchus ; - Rrochs - - - r
arcticus Fahr. 80-188 [26]   Greenl. Lapl.

Sciopithes Horn 76-62 [27]

* obscurus Horn 76-63   B.C.-Cal.
  significans Csy. 88-255 [27]
  brumalis Csy. 88-256 [27]
  angustulus Csy. 88-257 [27]
  arcuatus Csy. 88-257   Cal.
  intermedius VanD. 35-90   Cal.
  insularis VanD. 35-90   Cal.
  sordidus VanD. 35-91   Cal.
  setosus Csy. 88-258   Cal.

Myllocerus Schönh. 26-178 [28]
castaneus Rœlofs 73-168 [29]   Jap. Sib.-
              [?N.A.

Trachyphlœus
bifoveolatus Beck 17-22 [30]   Eur. N.S.
              [N.B. N.Y.
  davisi of authors (not Blatch.)[30]

Paraptochus Seidl. 68-35 [11]
* sellatus Boh. 59-126   Cal.
  setiferus VanD. 35-93   Cal.
  uniformis VanD. 35-95   Cal.

Stenoptochus
vanduzeei VanD. 35-95   Cal.

Omias
albus VanD. 35-96   Wash.

Anchitelus VanD. 36-19 [31]
alboviridis VanD. 36-19   Cal.

Peritelinus Csy. 88-263
* erinaceus VanD. 36-20   Cal.
  variegatus Csy. 88-263   Cal.
  oregonus VanD. 36-21   Ore.

[10] Van Dyke—34.
[11] Revision of genus, Van Dyke—35.
[20] Revision of genus, Buchanan—39.
[21] First reported in genus Naupactus, Watson—37, etc.
[22] Von Dalla Torre & Van Emden—36.

[23] Subfamily Brachyrhininæ, Van Dyke—36.
[24] Tribe Trigonoscutini, Van Dyke—36.
[25] Tribe Calyptillini, Van Dyke—36.
[26] Henriksen—35.
[27] Revision of genus, Van Dyke—35.
[28] To follow Neoptochus.
[29] Authority not known.
[30] Buchanan—37.
[31] To follow Periteloides in tribe Simoini, Van Dyke—36.

Nemocestes VanD. 36-22 [33]
* (*Geoderces* Horn 76-70)[22]
  horni VanD. 36-25   Mich. Kan. N.Y.
  melanothrix of Horn 76-71 [22]
  incomptus Horn 76-72   B.C.-Cal. Wyo.
  longulus VanD. 36-26   Cal.
  sordidus VanD. 36-26   Cal.
  montanus VanD. 36-27   Cal.-Wash.
  puncticollis Csy. 88-264   Cal.
  tuberculatus VanD. 36-28   Cal.
  kœbelei VanD. 36-183   Cal.
  kœbeli VanD. 36-29
  expansus VanD. 38-4   Cal.

Aragnomus
  setosus VanD. 36-30   Cal.

Eucyllus Horn 76-74 [33]
* (*Encyllus* VanD. 36-31)[33]
  vagans Horn 76-74   Ariz. Cal.
  echinus VanD. 36-31   Cal.
  unicolor VanD. 36-32   Cal.

Sitona
  cylindricollis Fahr. 40-269 [34][35]   Eur.
                           [Vt. Ont.
  lineatus Linn. 58-385 [36]   Eur. B.C.
  cockerelli Blais. 38-31
           [SanMiguel Id. L.Cal.

Lepidophorus Kby. 37-201 [17]
* (*Lophalophus* Lec. 76-120)[17]
  inquinatus Mannh. 52-351   Alaska
  rainieri VanD. 30-149   Wash.
  angulatus Buch. 36-4   Ida. Wash.
  plumosus Buch. 36-6   Colo.
  lineaticollis Kby. 37-201 Y.T. Can. Alas.
  pumilus Buch. 36-7   B.C.
  bakeri Buch. 36-8   Wash.
    s.utensis Buch. 36-10   Utah
  alternatus VanD. 30-150   Wash. Ore.
  setiger Hamilt. 95-347 N.Y.-Va. St.Vinc.

Lepyrus [37]
  labradorensis Blair 33-96   Arctic N.A.
  colon of Harring. [38]
  nordenskiöldi Faust. 85-34
    s.cæsius Csiki 34-7
      cinereus || VanD. 28-55 [39]

Listroderes
  obliquus Klug 29-?   S.A. Miss. Cal.
  obliquus Gyll. 34-277 [39][40]   [Afr.
  novica French 08-? [39][40] [Australia
  costirostris of authors [39]

Hylobius Germ. 17-340 [41]
* congener DallaT., Sch. & Marsh. 32-
                     [15 Mass.-Alas.
  confusus || Kby. 37-196 [41]
  pales Hbst. 97-31 Can.-Fla.-Tex.-Minn.
  radicis Buch. 34-252   N.Y. Minn.

Cholus Germ. 24-212 [42]
* catoleucus Chev. 81-482   N.A.
  cattleyæ Champ. 16-201 S.A. Wis. D.C.
  cattleyarum Barber 17-178 [42] [N.J.
  forbesi Pasc. 76-xxx   S.A. N.J.

Smicronyx
  dietzi Klima 34-95   N.J. D.C. Ind.
  nebulosus || Dietz 94-157 [42]
  fallaciosus Klima 34-95   Tex. Kan.
  fallax || Dietz 94-157 [42]
  albidosquamosus Klima 34-94   Ariz.
  albosquamosus || Dietz 94-168 [42]

Sthereus Mots. 45-374 [17]
* (*Trachodes* of authors, not Grm.)[17]
  ptinoides Germ. 24-327   B.C.-Alas.
  quadrituberculatus Mots. 52-355   Cal.-
                            [Alas.
  multituberculatus Buch. 36-179   Ore.-
                            [Alas.

Lobosoma Buch. 36-180 [17]
* (*Aparapion* of authors)[17]
  horridum Mannh. 52-354   Ore.-Alas.

Gastrotaphrus Buch. 36-180 [44]
* barberi Buch. 36-181   Cal.-B.C.

Thysanocnemis
  caseyi Klima 34-3   Nebr.
  brevis || Csy. 10-129 [43]

Magdalis
  piceæ Buch. 34-85   Mass. N.H.

Anthonomus
  blatchleyi Schenk. & Marsh. 34-40 Fla.
  australis || Blatch. 25-98 [45]
  17207, for 43-232 read 33-232 [46]
  univestus Schenk. & Marsh. 34-51 Fla.
  uniformis || Blatch. 16-300 [45]

Epimechus
  arenicolor Fall 01-265   Ariz.
  baccharidis Pierce 08-178 (Anthono-
                          [mus)[47]

Rhynchænus Clair.-Sch. 98-70
  (*Orchestes* Ill. 98-498)[48]

---

[17] Revision of genus, Buchanan—36.
[18] Buchanan—37.
[22] Revision of genus, Van Dyke—36.
[33] Revision of genus, Van Dyke—36.
[34] Buchanan in litt.
[35] Cæsar—36.
[36] Downes—38.
[37] Belongs in Cleoninæ according to Marshall—32 (p. 344).
[38] Brown—37.
[39] Csiki—34.
[40] Placed as synonyms of **costirostris** by Essig—33.

[37] Revision of genus, Buchanan—36.
[41] Revision of genus, Buchanan—34.
[42] Revision of genus, Klima—36.
[43] Klima—34.
[44] To follow **Lobosoma** in Trachodini, Buchanan —36.
[45] Schenkling & Marshall—34.
[46] Chamberlin in litt.
[47] Fall—34.
[48] Buchanan—39.

Gymnætron Schönh. 26-519 [49]
* antirrhini Payk. 00-257   Mass.-N.J.
   netum Germ. 21-307   Conn.-Va.-Ia.
   teter Fabr. 01-448 Can.-Ga.-Tex.-Wash.

Cleonus Schönh. 26-145
   DINOCLEUS Csy. 91-176
   capillosus Csiki 34-66   Cal.
      pilosus || Lec. 76-145 [50]
   structor Csiki 34-67   Cal. Ariz.
      molitor || Lec. 58-78 [50]
      CLEONIDIUS Csy. 91-176
      (*Apleurus* Chev. 73-78. part [50])
   americanus Csiki 34-64   Cal.
      basalis || Fall 97-242 [50]
   placidus Csiki 34-65   Cal.
      pacificus Fall 01-260 [50]
   coloradensis Csiki 34-64   Colo.
      canescens || Lec. 76-151 [50]
   lecontelus Csiki 34-64
      carinicollis || Lec. 76-152 [50]
   stratus Csiki 34-65   Colo.
      sparsus || Lec. 76-152 [50]

Lixus
   blatchleyi Csiki 34-120   Mich.
      cavicollis || Blatch. 22-113 [50]

Baris
   caseyi Hust. 38-51   Colo. Tex. La.
      vagans || Csy. 20-318 [51]
   blatchleyi Hust. 38-51   Fla.
      australis || Blatch. 20-168 [51]
   lengi Hust. 38-54   Ont.
      carbonaria || Bl. & Leng 16-356 [51]

Cosmobaris Csy. 20-344 [52]
* scolopacea Germ. 24-202 [52] [58] R.I.-Ia.
   americana Csy. 20-344 [53] [54]   [Eur.
      squamiger Hayes 36-27 [53] [54]
      sionilli Hayes 36-28 [53] [54]

Orchidophilus Buch. 35-45 [55]
* (*Acythopeus* of authors) [55]
   peregrinator Buch. 35-46   Hawaii, P.I.
                                    [D.C.
   gilvonotatus Barber 17-17 Orient, D.C.
                                    [Cal.
   orchivora Blackb. 00-61   N.J.

(*Acanthoscelis* = *Acanthoscelidus* Hustache 30-19
   utahensis Tanner 34-48   ·   Utah

Ceutorhynchus
   assimilis Payk. 00-257 [54]   Eur. Wash.
   punctiger Gyll. 37-538   Eur. Que.-N.J.-
      marginatus of authors [39]   [Ind.
   americanus Buch. 37-205   Que.-Md.-
      cyanipennis of authors [30] [Tex.-B.C.

Perigaster
   lituratus Dietz 96-457   Ont.- N.J.-Ia.
      longirostris Buch. 31-323 [30.] [Wash.

Phytobius Schönh. 36-458 [49]
*   (*Pelonomus* Thoms. 59-138) [49]
   17827 to 17834, belong here

   (*Eubrychius Thoms.*, *not No. Amer.*) [30]

Litodactylus Redt. 49-399
* griseomicans Schw. 91-165 [49] Wis.-Kan.-
                                    [Wash.

Eubrychiopsis Dietz 96-474 [49]
* lecontei Dietz 96-475   Mich. Wis. Ia.
   albertanus Brown 32-11   B.C.-Alb.

Collabismodes Champ. 05-541 [55]
* cubæ Boh. 44-342 [55]   Cuba, Fla.
   ater Blatch. 28-238 [57]   Fla.

Conotrachelus Schönh. 37-392
   (*Loceptes* Csy. 10-130) [30]
   recessus Csy. 10-130   Kan. Ckla. .Ark.
      atokanus Fall 13-65 [30]   [Tex.

Euscepes
   postfasciatus Fairm. 49-513   (Crypto-
                  [rhynchus) [48]   Cal. Tropics
      batatæ Waterh. 49-LXIX [48]

Sternochetus Pierce 17-143
   (*Cryptorhynchidius* Pierce 19-25) [48]
   mangiferæ Fahr. 75-139 (Cryptorhyn-
                  [chus) [48]   E. Ind. Fla.
   lapathi Linn. 46-591 (Cryptorhynchus) [48]
                                    [Eur. N.A.

Cossonus [58]
   americanus Buch. 36-112 ·   Que.-Cal.-
                                    [Wash.
      subareatus of authors (part) [57]
   pacificus VanD. 16-74   Ariz.-Cal.-
                                    [B.C. Que.
   rufipennis Buch. 36-114 Kans. Dak. Mo.
      subareatus of authors (part) [57]

Sitophilus Schönh. 38-979
   (*Calandra* of authors) [59]
   (*Calendra* of authors) [59]
   dietrichi Satt. 33-210   Miss. Ala.
   lucedalensis Satt. 33-212   Miss.

PLATYPODIDÆ

Platypus
   blanchardi Chap. 66-185 [60]   Fla.
   18168, for 15-48 read 16-97 [46]

[49] Revision of genus, Buchanan—37.
[50] Csiki—34.
[51] Hustache—38.
[52] Revision of genus, Hayes—36.
[53] Fall—37.
[54] Listed as valid species by Hayes—36.
[55] Revision of genus, Buchanan—35
[56] Baker—36.
[30] Buchanan—37.
[46] Chamberlin in litt.
[48] Buchanan—39.
[57] Junk Cat. p. 151, p. 210.
[58] Revision of pacificus-group, Buchanan—36.
[59] Satterthwait in litt.
[60] Valid species, Schedl—33.

SCOLYTIDÆ

Scolytus Geoffr. 62-310 [61]
* (*Eccoptogaster* Hhst. 93-124) [61]
rugulosus Ratz. 37-187   Eur. N.A.
muticus Say 24-323   E.U.S.
fagi Walsh 67-38   Ill. Tex.
sulcatus Lec. 68-167   Conn.-N.J.
reflexus Blackm. 34-13   Ariz.
wickhami Blackm. 34-13   Colo. N.M.
    [Ar. Ut.
tsugæ Sw. 17-32   Cal.-B.C.-Alb.
monticolæ Sw. 17-32   Or.-B.C.-Wyo.
quadrispinosus Say 24-323   N.A.
    caryæ Riley 67-? [61]
subscaber Lec. 76-371   Cal.-B.C.
oregoni Blackm. 34-18   Ore.
robustus Blackm. 34-19   Colo. N.M.
    [Ar. Ut.
præceps Lec. 76-371   Cal.-Tex.
opacus Blackm. 34-20   Colo. Ut. Mont.
abietis Blackm. 34-21   Ida.
ventralis Lec. 68-167   B.C.-Cal.-
    [N.M.-Mont.
scolytus Fahr. 75-59   Eur. ?U.S.
    californicus Lec. 68-165 [61]
sobrinus Blackm. 34-23 Wash. Cr. Wyo.
laricis Blackm. 34-24   Ida. Mont.
fiskei Blackm. 34-25   N.M. Colo.
unispinosus Lec. 76-371   Cal.-B.C.
piceæ Sw. 10-34   Que.-N.Y.-Mont.
multistriatus Marsh. 02-54 Eur. Mass.-
    [Pa.

Hylesininæ [62]

Crypturgus [62]

Dolurgus [62]

Polygraphus [62]

Carphoborus [62]
cressatyi Bruck 36-36   Cal.
vandykei Bruck 33-104   Cal.
swainei Bruck 33-105   Cal.

Renocis [63] [63]
penicillatus Brown 33-239   Cal.

Pseudocryphalus
maclayi Bruck 36-35   Cal.

Chramesus Lec. 68-168 [63] [63]
* (*Rhopalopleurus* Chap. 69-46) [63]
hicoriæ Lec. 68-168   Can.-Ga.-Miss.
icoriæ Lec. 76-375 [63]
lecontei Chap. 69-46 [63]
asperatus Schffr. 08-220   Ariz.
subopacus Schffr. 08-221   Tex. Ariz.
canus Blackm. 38-541   La.
gibber Blackm. 38-541   N.Mex.
chapuisii Lec. 76-375 Pa.-Fla.-Tx.-Kan.
dentatus Schffr. 08-221   Ariz.
mimosæ Blackm. 38-544   Tex. Mex.

Phthorophlœus [62]

Tomicus [62]

Dendroctonus [62]

Phlœosinus [62]
piceæ Sw. 34-205   Que.
granulatus Bruck 36-33   Cal.
setosus Bruck 33-54   Cal.
frontalis Bruck 33-55   Cal.
swainei Bruck 33-56
    minutus || Sw. 17-9 [64]

Chætophlœus [62]

Xylechinus [62]

Leperisinus [62]

Dendrosinus [62]

Scierus [62]

Hylastinus [62]

Alniphagus [62]

Hylurgopinus [62]

Pseudohylesinus [62]
serratus Bruck 36-37   Cal.

Hylurgops [62]

Hylastes [62]
porculus Er. 36-49   E.N.A.
    granosus Chap. 69-73 [65]
    scaber Sw. 17-18 [65]
swainei Egg. 34-25   E.Can. Minn.
    porculus of Sw. 18-78 [65]

---

[61] Revision of genus. Blackman—34.
[62] Rearrangement of genera in subfamily Hylesininæ, Bruck—36.
[63] Revision of genus. Blackman—38.
[64] Bruck—33.
[65] Eggers—34.

Pseudothysanoes Blackm. 20-46 [46]

* hopkinsi Blackm. 28-200		Cal.
partoni Bruck 36-32		Cal.
drakei Blackm. 20-48		N.Y.
rigidus Lec. 76-362	Can.	Mich.
sulcatus Bruck 36-33		Cal.
phorodendri Blackm. 28-202	Tex.	Ariz.
lecontei Blackm. 20-49		D.C.
sedulus Blackm. 28-204		Ariz.
gambetti Blackm. 28-205	Ariz.	N.M.
barberi Blackm. 28-206		Ariz.

Cactopinus Sz. 99-11 [48]

* hubbardi Sz. 99-11	Ariz.
pini Blackm. 38-153	Cal.
rhois Blackm. 38-154	Cal.
kœhelei Blackm. 38-156	Cal.

Pterocyclon Eich. 68-276 [47]

obliquecaudatum Schedl 35-351	Cal.

Hypothenemus

citri Ebeling 35-21	Cal.

Cryphalus

18374, for Cal. read Ore.[66]
18375, for Cal. read Ore.[68]

Gnathotrichus

22039, for Colo. read B.C. Wash.[69]
22040, for B.C. Wash. Ore. Cal. read
[S.D. Colo.

Ancyloderes Blackm. 38-205 [70]

* pilosus Lec. 68-154	Cal.
pilosulus Lec. 68-166 [71]	
saltoni Blackm. 38-206	Ariz.

Ips

montanus Eichh. 81-219 [72]	Cal. or Tex.

Xyleborinus

tsugæ Sw. 34-204	B.C.
librocedri Sw. 34-205	Ore.

[66] Chamberlin in litt.
[69] H. B. Leech in litt.
[70] To follow Pseudopityophthorus.
[71] Blackman—38.
[72] Eggers—34.

[46] Revision of genus, Bruck—36.
[47] Used as valid genus by Schedl—35.

ADDITIONAL NECROLOGY

Bethune, C. J. S., d. April 18, 1932.
Beutenmuller, W., d. February 24, 1934.
Bolster, P. G., d. May 22, 1932.
Carr, F. S. b. January 1, 1881; d. May 16, 1934.
Fenyes, A., b. November 17, 1863; d. February 22, 1937.
Field, G. H., d. June 12, 1937.
Handlirsch A., d. August 28, 1935.
Horn, W., d. July 10, 1939.
Johnson, C. W., d. July 19, 1932.
Kellogg, V. L., b. December 1, 1867; d. August 8, 1937.
Klages H. G., d. October 26, 1936.
Nevermann, F., d. July 3, 1938.
Psota, F. J., d. April 13, 1936.
Stickney, F. S., b. August 6, 1892; d. August 15 1936.
Tillyard R. J., d. January 13, 1937.
Urich, F. W., d. June 22, 1937.
Wheeler, W. M., d. April 19, 1937.
Wickham, H. F., b. October 26, 1866; d. November 17, 1933.

As this supplement goes to press we have word of the passing, on November 14th, 1939, of our most eminent worker on North American Coleoptera,

HENRY CLINTON FALL.

The publisher, his close friend for fifty years, and his many entomological and other friends feel their loss very deeply, and workers on Coleoptera all over the land will sadly miss his assistance and encouragement.

In spite of severe physical handicaps, patiently endured throughout his lifetime, Dr. Fall accomplished a vast amount of most valuable work on the taxonomy of North American beetles, describing some 1,400 new species—few of which were ever challenged as being synonyms— and his excellent monographic works on various families and genera of these insects are well known and in constant use.

BIBLIOGRAPHY

Abeille de Perrin, E.
97. Notes sur les Buprestides paléarctiques<Rev. d'Ent., 16, 1-33.

Alsterlund, J. F.
37. Larva of Chalcodermus collaris   <Proc.Ent.Soc.Wash.,39, 216-222.

Allard, E.
57. Desc...n.sp. of Lithocharis   <Ann.Soc.Ent.Fr.,(3),5,747-748.

Andersoh, O.
97. In Hoppe, Ent. Tasch. f.1797.   Regensburg,1797,252 pp.

Anderson, W. H.
36. Comp. study of labium of col. larvæ   <Smiths.Misc.Coll.,95,no.13,29pp.
38. Desc. of larvæ of Chætocnema   <Proc.Ent.Soc.Wash.,40,161-169.

Andre, F.
37. Brood A june beetles in Iowa   <Ia.St.Coll.Jl.Sci.,11,267-280

Arrow, G. J.
11. Notes on...Dynastinæ, with desc.   <Ann.Mag.Nat.Hist.,(8),8,151-176.
13. Notes on Lamell. Col. of Japan...   <Ann.Mag.Nat.Hist.,(8),12,394-408.
27. Note on...Aserica   <Proc.Ent.Soc.Wash.,29,69-70.
33. Further note on...Aserica   <Proc.Ent.Soc.Wash.,35,71-73.
35. Contrib. to class...Lucanidæ   <Tr.R.Ent.Soc.Lond.,83,105-125.
37. Scarabæidæ: Dynastinæ   <Junk Col.Cat.,pars 156.
37. Syst. notes...Dynastinæ   <Tr.R.Ent.Soc.Lond.,86,35-57.

Aubé, C.
50. Desc...Col...Eur. et Algérie   <Ann.Soc.Ent.Fr.,(2),8,299-346.

Back, E. A. and Cotton, R. T.
37. ...(Anthrenus vorax)...in U.S.   <Proc.Ent.Soc.Wash.,38,1936(1937), 189-198.

Baker, W. W.
36. Notes on...Ceutorhynchus assimilis   <Can.Ent.,68, 191-193.

Balch, R. E.
37. Notes on...(Nacerdes melanura)   <Can.Ent.,69,1-5

Balduf, W. V.
35. Bionomics of entomophagous Col.   New York, 1935, 220 pp.

Balfour-Browne, F.
34. Proventriculus in Dytiscidæ   <Proc.R.Ent.Soc.Lond.,3,241-244.
34. Syst. notes on Brit. aquatic Col.   <Ent.Mo.Mag.,70,127-132, 146-150, 175-180, 224-230, 247-255.
35. Syst. notes on Brit. aquatic Col.   <Ent.Mo.Mag.,71, 7-14, 195-201, 219-226, 246-249.
36. Syst. notes on Brit. aquatic Col.   <Ent.Mo.Mag.,72, 28-31, 68-77, 121-126.

Ballou, C. A. and Siepmann, C. G.
39. Hister ciliatus from Ariz.   <Bull.Br.Ent.Soc.,33,1938 (1939), 242-243.

Bänninger, M.
25. 9er Beitrag...Carabinæ: Nebriini   <Ent.Mitt.,14,180-195, 256-281, 329-343.
33. Über Carabinæ...   <Deutsche Ent.Zeit.,1933,81-124.
37. Monog. subtr. Scaritina I.   <Deutsche Ent.Zeit.,1937,81-160.

Barber, H. S.
37. Some spp. of Colaspis near brunnea   <Proc.Ent.Soc.Wash.,38,1936(1937), 198-204.

Barrett, R. E.
33. N.sp. of N.A. Scarabæidæ   <Can.Ent.,65,129-132.
35. N.sp. of N.A. Scarabæidæ, II.   <Can.Ent.,67, 49-52.

Basilewsky, P.
37. Desc...Carabus...Alaska   <Bull.Soc.Ent.Fr.,42, p.63.

Bates, F.
04. Rev. of Pelidnotinæ   <Tr.Ent.Soc.Lond.,1904,249-276.

Bates, H. W.
69. N.spp. of Col. from Nicaragua   <Tr.Ent.Soc.Lond.,1869,383-389.

Baudi di Selve, F. and **Truqui,** E.
48. Studi Entomologici, I.   Torino, 1848, 376 pp.
Beaulne, J.-L.
36. Cont...Col. du Canada   <Le Nat.Canad.,63,158-164.
Bedard, W. D.
38. Annot. list of ins. of Douglas fir   <Can.Ent.,70,188-197.
Bedel, L.
21. Faune des Col. du Bassin de la Seine   Paris, IV, 2, 1921.
Belkin, J. N.
34. Addit. to N.Y. St. list of **Ins.**   <Bull.Br.Ent.Soc.,28, 1933 (1934), 220-222.
Benedict, W.
34. Another Cicindela   <Pan-Pac.Ent.,10, p. 76.
Benesh, B.
32. Notes on some stag-beetles   <Ent.News, 43, 40-41.
37. Some notes on bor. Am. Dorcinæ   <Tr.Amer.Ent.Soc.,63,1-16.
39. Desc. n. forms Pseudolucanus capreolus<Ent.News, 49, 1938(1939),271-274.
Berg, C.
81. Ent...Indianergeb. der Pampa   <Stett.Ent.Zeit.,42,36-72.
81. Rev. Argent. Cantharis   <Stett.Ent.Zeit.,42,301-309.
01. Silfidos argentinos   <Com.Mus.Nac.Buenos Aires,1, (9),325-330.
Berlioz, J.
33. Note sur... Desmocerus piperi   <Bull.Mus.Natl.Hist.Nat.Paris,
Bernhauer, M.
08. Beitr...paläark. Staphyliniden   <Münch.Kol.Zeit.,3,320-335.
08. 14 Folge n. Staph. der paläark. **Fna.**   <Verh.zool.bot.Ges.Wien,58,32-41.
15. Beitr...paläark. Staphyl. II.   <Münch.Kol.Zeit.,4,1914 (1915), 1-10, 33-45.
29. Neue Staphyl. aus Mittelamerika   <Wiener Ent.Zeit.,46,186-208. (2),5,111-113.
Bierig. A.
34. A new N.A. Astenus   <Mem.Soc.CubanaHist.Nat.,8,29-30.
34. Neues...Cafius...   <Rev. de Ent.,4,65-70.
38. Sobre...Acylophorus...   <Mem.Soc.Cubana Hist.Nat.,12, 119-138.
38. Desc...generos...Staphylinidæ   <Mem.Soc.Cubana Hist.Nat.,12, 139-147.
Blackburn, T.
04. Rev. Australian Aphodiides...   <Proc.Soc.Victoria,17,145-181.
Blackburn, T. and Sharp, D.
85. Mem. on Col. of Hawaiian Is.   <Tr.R.Dublin Soc.(2),3,119-196.
Blackman, M. W.
34. Rev. study of Scolytus in N.A.   <U.S.Dpt.Agr.,Tech.Bull.431,30 pp.
38. Genus Chramesus in N.A.   <Jl.Wash.Ac.Sci.,28,534-545.
38. N.spp. of Cactopinus   <Proc.Ent.Soc.Wash.,40,151-157.
38. Ancyloderes, n.gen. of Scolytidæ   <Proc.Ent.Soc.Wash.,40,204-206.
Blackwelder, R. E.
34. Prostheca or mandibular appendage   <Pan-Pac.Ent.,10,111-113.
36. Morphology of Staphylinidæ   <Smiths.Misc.Coll.,94,no.13,102 pp.
36. Rev. N.A. Tachyporus   <Proc.U.S.Nat.Mus.,84,39-54.
38. Rev. N.A. Coproporus   <Proc.U.S.Nat.Mus.,86,1-10.
39. Generic rev. of Pæderini   <Proc.U.S.Nat.Mus.,87, 93-125.
Blair, K. G.
33. Col. coll. on Akpatok Id.   <Ann.Mag.Nat.Hist.,(10),12,93-96.
34. Beetle larvæ   <Proc. & Tr.So.Lond.Ent.& Nat. Hist.Soc.,1933-34, 89-110.
Blaisdell, F. E.
33. N.sp. Helops from Guadalupe Id.   <Pan-Pac.Ent.,9,88-90.
33. Note on Euschides cressoni   <Pan-Pac.Ent.,9.p.152.
33. Studies in Tenebrionidæ, III.   <Tr.Amer.Ent.Soc.,59.191-210.
33. Monogr. rev. of Centronopus   <Tr.Amer.Ent.Soc.,59.211-228.
34. N.sp. of Vectura from So.Cal.   <Pan-Pac.Ent.,10,71-73.
34. Note on Apsena barbaræ   <Pan-Pac.Ent.,10,p.110.

34. N.sp. of Hoppingiana from B.C. $<$Can.Ent.,66,150-152.
34. Second. sex. char. in Phlœodes $<$Pan-Pac.Ent.,10, p.110.
34. Stud. in Corticeus (Hypophlœus) $<$Ent.News, 45,187-191.
34. Studies in Auchmobius $<$Tr.Amer.Ent.Soc.,60,223-264.
34. Rare N.A. Coleoptera $<$Tr.Amer.Ent.Soc.,60,317-326.
35. Two n.spp. of Eleodes $<$Can.Ent.,67,28-31.
35. Facts from rearing Coniontis $<$Ent.News, 46, 119-123.
35. A n. Triorophid from Death Valley $<$Pan-Pac.Ent.,11,125-129.
36. Monog. Rev. of Stibia $<$Tr.Amer.Ent.Soc.,62,57-105.
36. Studies in Melyridæ, 11. $<$Pan-Pac.Ent.,12,184-190.
36. Notes on Eleodes letcheri & rileyi $<$Pan-Pac.Ent.,12, p.183.
36. Notes on Eschatomoxys wagneri $<$Pan-Pac.Ent.,12, p.120.
36. Two n.spp. of Notoxus $<$Can.Ent.,68, 144-148.
36. Facts on rearing Tenebrionidæ $<$Ent.News, 47, p.39.
36. Two n.spp. of Euschides $<$Tr.Amer.Ent.Soc.,62,223-230.
37. Third n.sp. of Centronopus from Cal. $<$Pan-Pac.Ent.,13,95-96.
37. Misc. studies in Coleoptera, 5. $<$Tr.Amer.Ent.Soc.,63,127-145.
37. N.sp. of Cryptophagus... $<$Ent.News, 48, 158-160.
38. Gen. syn. of tribe Dasytini $<$Tr.Amer.Ent.Soc.,64,1-31.
38. N.sp. of Sitona from San Miguel Id. $<$Pan-Pac.Ent.,14,31-33.
38. N.sp. of Listrus from Calif. $<$Pan-Pac.Ent.,14,165-167.
39. Hispinæ of genus Stenopodius $<$Tr.Amer.Ent.Soc.,64, 1938 (1939), 421-447.

39. Relationships of 'subfam. & tribes of $<$Tr.Amer.Ent.Soc.,65,43-60. Teneb.

Blake, D. H.
33. Rev. Disonycha of N.A. $<$Proc.U.S.Nat.Mus.,82,no.28,66 pp.
33. Two n.sp. of Systena, with notes $<$Proc.Ent.Soc.Wash.,35,180-183.
35. Notes on Systena $<$Bull.Br.Ent.Soc.,30,89-108.
36. Altica bimarginata... $<$Proc.Ent.Soc.Wash.,38,13-24.
36. Redispos. of Monoxia puncticollis, etc. $<$Jl.Wash.Ac.Sci.,26,423-430.
37. A new Monoxia from L.Cal. $<$Zoologica,22,89-91.

Blatchley, W. S.
36. Change of name in Staphylinidæ $<$Ent.News,47,255-256.
38. Coll. beetles in Florida $<$Ward's Ent.Bull.,6, nos.1 & 2.

Bleasdell, C. G.
37. Rhynchophora of Iowa $<$Ia.St.Coll.Jl.Sci.,11,405-445.

Blocker, J. C. von
36. Status of Phyllophaga lanceolata $<$Bull.So.Cal.Ac.Sci.,35,52-61.
37. Status of Phyllophaga cribrosa $<$Bull.So.Cal.Ac.Sci.,36,83-88.

Blood, R.
35. Anatomy of Pyrota mylabrina $<$Jl.N.Y.Ent.Soc.,43,1-17.

Boheman, C. H.
40. in Schönherr, 1840.
44. in Schönherr, 1844.

Borchmann, F.
36. Lagriidæ $<$Gen.Ins.,fasc.204, 561 pp.

Bottimer, L. J.
35. New Acanthoscelides from e.U.S. $<$Ent.News, 46,127-129.
36. Bruchus brachialis in Georgia $<$Jl.Econ.Ent.,29,p.807.
37. Notes on Bruchus brachialis $<$Jl.Econ.Ent.,30,p. 379.

Böving, A. G.
33. Desc. of larva of Decadiomus pictus $<$Proc.Biol.Soc.Wash.,46,101-104.
36. Desc. of larva of Plectris aliena $<$Proc.Ent.Soc.Wash.,38,169-185.

Bowman, J. R.
34. Pselaphidæ of N. A.   Pittsburgh, 1934, 149 pp.

Breuning, S.
32. Monographie der Gattung Carabus $<$Best.-Tab.eur.Col.,104,1-288.
32. Monographie der Gattung Carabus $<$Best.-Tab.eur.Col.,105,289-496.
33. Monographie der Gattung Carabus $<$Best.-Tab.eur.Col.,106,497-704.
33. Monographie der Gattung Carabus $<$Best.-Tab.eur.Col.,107,705-912.
34. Monographie der Gattung Carabus $<$Best.-Tab.eur.Col.,108,913-1120.
35. Monographie der Gattung Carabus $<$Best.-Tab.eur.Col.,109,1121-1360.
37. Monographie der Gattung Carabus $<$Best.-Tab.eur.Col.,110,1361-1610.

Bridwell, J. C. and Bottimer, L. J.
  33. Bruchus brachialis in U.S.   <Jl.Agric.Res.,46,739-751.

Brown, A. W. A.
  34. Contr. to insect fauna of Timagini   <Can.Ent.,66,206-211, 220-231,
                              242-252, 261-267.

Brown, W. J.
  33. N.sp. of Coleoptera, IV   <Can.Ent.,65,43-47.
  33. Studies in the Elateridæ, I   <Can.Ent.,65,133-141.
  33. Studies in the Elateridæ, II   <Can.Ent.,65,173-182.
  33. Two n.sp. of Silphidæ   <Can.Ent.,65,213-215.
  34. N.sp. of Coleoptera, V   <Can.Ent.,66,22-24.
  34. Entomological Record, Coleoptera   <25th & 26th Rpt.Que.Soc.Prot.
                              Plants,145-151.

  34. Amer. spp. of Dalopius   <Can.Ent.,66,30-39, 66-72, 87-96,
                              102-110
  35. Ludius: cruciatus & edwardsi groups <Can.Ent.,67,1-8.
  35. Ludius: æripennis group   <Can.Ent.,67,125-135.
  35. Ludius: cribrosus & volitans groups <Can.Ent.,67,213-221.
  36. Ludius: semivittatus & nitidulus groups<Can.Ent.,68,11-20.
  36. Ludius: fallax & triundulatus groups <Can.Ent.,68,99-107.
  36. Entomological Record, Coleoptera   <27th Rpt.Que.Soc.Prot.Plants,
                              1934-35 (1936), 90-92.
  36. Ludius: inflatus. group   <Can.Ent.,68,133-136.
  36. Changes of names   <Can.Ent.,68,p.142.
  36. Ludius: propola group   <Can.Ent.,68,177-187.
  36. Notes on Elateridæ   <Can.Ent.,68,246-252.
  37. Col. of Canada's e.Arctic   <Can.Ent.,69,106-111.
  37. Desc. of gen. & spp. of Leiodidæ   <Can.Ent.,69.158-165, 170-174.
  37. N.A. spp. of Anisotoma   <Can.Ent.,69, 193-203.
  38. Some n. Canadian Chrysomelidæ   <Can.Ent.,70,35-38.

Bruck, C. R.
  33. Two n.spp. of Phlœosinus   <Can.Ent.,65,54-56.
  33. N.spp. of Carphoborus with key   <Can.Ent.,65,103-106.
  33. New bark beetle from So.Cal.   <Can.Ent.,65,239-240.
  36. New Scolytidæ from So.Cal.   <Bull.So.Cal.Ac.Sci.,35,30-38.
  36. Syn. rev. Hylesininæ of w.N.A.   <Bull.So.Cal.Ac.Sci.,35,38-51.

Brullé, G. A.
  40. in d'Orbigny, Voy. Amer. mérid...   Paris,1837-43 vol.6.pt.2.

Buchanan, L. L.
  34. New N.A. Magdalis from blue spruce <Proc.Ent.Soc.Wash.,36,85-87.
  34. Henry Frederick Wickham   <Proc.Ent.Soc.Wash.,36,60-64.
  35. T. L. Casey & the Casey colln.   <Smiths.Misc.Coll.,44.no.1, 15 pp.
  35. N.g. & sp. of orchid weevils   <Proc.Haw.Ent.Soc.,9,45-48.
  35. Notes on Collabismodes cubæ   <Bull.Br.Ent.Soc.,30,125-126.
  35. N.sp. of N.A. Hylobius, with key   <Proc.Ent.Soc.Wash.,36,1934
                              (1935), 252-256.
  36. Synopsis of Lepidophorus   <Bull.Br.Ent.Soc.,31,1-10.
  36. Genus Panscopus   <Smiths.Misc.Coll.,94.no.16,18 pp.
  36. Pacificus group of Cossonus   <Pan-Pac.Ent.,12,111-116.
  36. Syst. notes on Trachodinæ   <Proc.Ent.Soc.Wash.,37,1935
                              (1936), 178-183.
  37. Notes on Curculionidæ   <Jl.Wash.Ac.Sci.,27,312-316.
  37. N.sp. of Ceutorhynchus   <Bull.Br.Ent.Soc.,32,205-207,.
  37. Nomencl. of Listroderes obliquus   <Proc.Ent.Soc.Wash.,38, 1936
                              (1937), 204-208.
  39. Change of names in Carab. & Rhynch. <Proc.Ent.Soc.Wash.,41,79-82.
  39. Spp. of Pantomorus of N.A.   <U.S.Dpt.Agr.,Misc.Pub.no.341,
                              39 pp.

Burmeister, H. and Schaum, H.
  40. Krit. Rev. der Lamell. Melitophila <Germar, Ztschr.Ent.,2,(2),353-420.
  41. Krit. Rev. der Lamell. Melitophila <Germar, Ztschr.Ent.,3,226-282.

Caesar, L.
  36. Notes...pest of sweet clover in Ont. <66th Ann.Rpt.Ent.Soc.Ont., 1935
                              (1936), 54-56.

Cameron, M.
    21. N.spp. of Staphylinidæ from Singapore <Tr.Ent.Soc.Lond.,1920 (1921), IV.   347-413.
    22. Desc. n. Staphyl. from W. I. II.   <Ann.Mag.Nat.Hist.,(9),9,113-128,
    33. Remarks on...Motschoulsky's...   <Ent.Mo.Mag.,69, 219-220.
    Staphylinidæ

Candeze, E.
    82. Elaterides nouveaux, III.   <Mem.Liege,(2),9,1-117.

Canova, M. F.
    36. Annot. list of Lepturini of Ore.   <Pan-Pac.Ent.,12,126-132.

Carpenter, G. D. H.
    38. Ins. coll. in w.Greenland   <Ann.Mag.Nat.Hist.,(11),1,529-553.

Carter, H. J. and Zeck, E. H.
    29. Monog. of Australian Dryopidæ   <Austr.Zool.,6,50-72.

Cartwright, O. L.
    34. New Atænius from Fla.   <Can.Ent.,66,200-201.
    34. List of Scarabs...S.C.   <Ent.News,45,237-240, 268-269.
    35. Tiger beetles of S.C.   <Bull.Br.Ent.Soc.,30,69-77.
    35. N.sp. of Phyllophaga from Fla.   <Ent.News, 46, 102-104.
    38. So.African Onthophagus in U.S.   <Ent.News, 49, 114-115.

Casey, T. L.
    09. Studies...Caraboidea & Lamellicornia <Can.Ent.,51,253-284.

Cazier, M. A.
    36. Notes on Cicindela plutonica   <Pan-Pac.Ent.,12,123-124.
    37. N.sp. of Valgus...   <Pan-Pac.Ent.,13,190-192.
    37. Rev. Pachydemini of N.A.   <Jl.Ent. & Zool.,29,73-87.
    37. Two new Calif. Cœnonycha   <Bull.So.Cal.Ac.Sci.,36,125-128.
    37. Four new Calif. Col.   <Pan-Pac.Ent.,13,115-118.
    37. A new Calif. Omus   <Pan-Pac.Ent.,13, p.94.
    37. Rev...groups of Cicindela   <Bull.So.Cal.Ac.Sci.,35,156-163.
    38. Two new Calif. Acmæodera   <Bull.So.Cal.Ac.Sci.,37,137-140.
    38. A new Calif. Polyphylla...   <Pan-Pac.Ent.,14,161-164.
    38. Gen. rev...N.A. Cremastocheilini   <Bull.So.Cal.Ac.Sci.,37,80-87.
    38. New Acmæodera & Chrysobothris   <Bull.So.Cal.Ac.Sci.,37,12-17.

Chagnon, G.
    33. Desc. du genre de vie...Coleop.   <Le Nat.Canad.,60,289-302.
    33. Contr...Col...Quebec [1]   <Le Nat.Canad.,60.166-178, 202-213, 289-302, 319-330, 343-351.

[1] Also published separately in 6 parts, 1934-1939.
    34. Lasiotrechus discus in N.A.   <Can.Ent.,66, p.168.
    34. Deleaster dichrous in N.A.   <Can.Ent.,66, p. 282.
    34. Contr...Col...Quebec [1]   <Le Nat.Canad.,61,18-26, 84-95, 99-110, 137-157, 182-198, 215-230, 269-282, 309-319.

    35. Col. in Polyporus betulinus   <Can.Ent.,67, p.278.
    35. Contr...Col...Quebec [1]   <Le Nat.Canad.,62,40-52, 130-141, 165-176, 222-233, 333-345.

    36. Col. du Champignon du Bouleau   <Le Nat.Canad.,63,31-32.
    36. Staphylinus globulifer in E.Can.   <Can.Ent.,68, p.116.
    36. Contr...Col...Quebec [1]   <Le Nat.Canad.,63, 104-112, 201-210, 241-251.

    36. Ins. nouv...de Lanoraie, Que.   <Le Nat.Canad.,63,164-166.
    36. Staphylinus globulifer dans Can.   <Le Nat.Canad.,63, p.265.
    37. Contr...Col...Quebec [1]   <Le Nat.Canad.,64,22-30, 101-117, 218-228, 243-253.

    38. Contr...Col...Quebec [1]   <Le Nat.Canad.,65, 13-23, 157-166.

Chamberlin, W. J.
    33. Syn. of Polycesta...   <Jl.N.Y.Ent.Soc.,41,37-46.
    34. N.spp. of Chrysobothris...   <Pan-Pac.Ent.,10,35-42.
    38. Six n.spp. of Chrysobothris...   <Pan.Pac.Ent.,14,10-16.
    38. New Buprestidæ from Calif.   <Jl.N.Y.Ent.Soc.,46,445-447.

Champion, G. C.
    13. Notes on C.A. Coleoptera...   <Tr.Ent.Soc.Lond.,1913,58-169.
    23. Some Indian Coleoptera   <Ent.Mo.Mag.,59,165-179.

Chapin, E. A.
32. Autoserica pro Aserica   <Proc.Ent.Soc.Wash.,34,122-124.
34. A new Scarab...Charleston, S.C.   <Proc.Biol.Soc.Wash.,47,33-35.
34. New Listrochelus...Rky.Mts.   <Proc.Biol.Soc.Wash.,47,93-94.
35. Rev. of Chlænobia   <Smiths.Misc.Coll..94,no.9, 20 pp.
38. Three Japanese...Serica   <Jl.Wash.Ac.Sci.,28,66-68.
38. Nomen. & tax. of Glaphyrinæ   <Proc.Biol.Soc.Wash.,51,79-86.

Chaudoir, M. de
50. Mem. sur la...Carabiques   <Bull.Moscou,23,II,349-460.
61. Mater...Cicindeletes & Carabiques   <Bull.Moscou,34,II,491-576.
62. Mater...á l'étude des Carabiques   <Bull.Moscou,35,IV,275-320.
69. Desc. de Calosoma nouveaux...   <Ann.Soc.Ent.Fr.,(4),9,367-378.

Chenu, J. C.
60. Encycl. d'Hist. Nat. Col.   Paris, 3 vols.,1851-1860.

Chevrolat, L. A. A.
33. Col. du Mexique   Strasbourg, 8 pts., 1833-1835.
39. Note sur...Ega...   <Rev.Zool.,1839, p. 308.
41. Col. du Mexique.   <Guerin,Mag.Zool.,11,nr.55-59.
43. Col. du Mexique.   <Rev.Mag.Zool.(2),5,(pp. 1-37).

Cooper, K. W.
33. Amer. spp. of Alobates   <Bull.Br.Ent.Soc.,28,105-107.
33. Xenorhipis brendeli from L.I.   <Bull.Br.Ent.Soc.,28,p.115.
33. N.sp. of Staphylinus   <Can.Ent.,65,264-265.
33. Micromalthus debilis in N.Y.   <Jl.N.Y.Ent.Soc.,41,545-546.
34. Micromalthus debilis in N.Y.   <Bull.Br.Ent.Soc.,29,p.130.
34. Taxonomy of Byrrhidæ   <Bull.Br.Ent.Soc.,29,130-131.
35. Suppl. to N.Y. St. list of Col.   <Bull.Br.Ent.Soc.,30,142-159.

Corporaal, J. B.
33. Further notes on Cleridæ   <Tijdschr.v.Ent.,76,115-118.

Couper, W.
65. Desc. of n.sp. of Canad. Col.   <Can.Nat.& Geol..2.no.2 (n.s.), 60-63.

Cresson, E. T.
35. Gibbium psylloides in Phila.   <Ent.News,46,p.230.

Criddle, N. and Handford, R. H.
33. Lema trilineata in Man.   <Can.Ent.,65,150-151.

Crowson, R. A.
38. Metendosternite in Col.   <Tr.R.Ent.Soc.Lond.,87,397-416.

Csiki, E.
33. Carabidæ: Harpalinæ VIII   <Junk Col.Cat.,pars 126.
33. Carabidæ: Carabinæ III   <Junk Col.Cat.,pars 127 (part)
34. Curculionidæ: Hyperinæ   <Junk Col.Cat.,pars 137.
34. Curculionidæ: Cleoninæ   <Junk Col.Cat.,pars 134.
36. Curcul.: Rhynchophorinæ & Cossoninæ <Junk Col.Cat.,pars 149.

von Dalla Torre, K. W., Schenkling, S., and Marshall, G. A. K.
32. Curcul: Pissodinæ   <Junk Col.Cat.,pars 125.

von Dalla Torre, K. W. and van Emden, M. & F.
36. Curcul.: Brachyderinæ I   <Junk Col,Cat.,pars 147.
37. Curcul.: Brachyderinæ II   <Junk Col.Cat..pars 153.

von Dalla Torre, K. W. and Voss, E.
35. Curcui.: Otidocephalinæ, Ithycerinæ, Belinæ, Petalochilinæ, Oxycoryninæ   <Junk Col.Cat..pars 144.
37. Curcul.: Mesoptiliinæ, Rhynchitinæ I   <Junk Col.Cat..pars 158.

Daniel, K.
03. Bestim.-Tab.eur.Kol. (Carabidæ)   <Münch.Kol.Zs.,1,155-173.

Darlington, P. J.
28. Ochthebius bruesi   <Psyche,35,p.3.
33. Cbit. of P. G. Bolster   <Psyche,40,87-88.
33. Subspp. of Sphæroderus canadensis   <Psyche,40,62-64.
33. N.tribe of Carabidæ from w.U.S.   <Pan-Pac.Ent.,9,110-114.
34. Subspp. of Chlænius leucoscelis   <Pan-Pac.Ent.,10,115-118.
35. Three W.I. Carabidæ in Fla.   <Psyche,42,p.161.

35. Megacephala angustata in **U.S.**   <Psyche,42,161-162.
36. Two introd. sp. of Amara   <Psyche,43,p.20.
36. Variation & atrophy...wings...Cara-<Ann.Ent.Soc.Amer.,29,136-179.
bidæ
36. Spp. of Stenomorphus...   <Pan-Pac.Ent.,12,33-44.
36. ...Pterostichus & Colpodes from Ariz. <Bull.Br.Ent.Soc.,31,150-153.
37. W.I. spp. of Osorius   <Bull.Mus.Comp.Zool.80,no.6, 283-
301.
38. Dytiscus habilis in Texas   <Psyche,45,p.84.
38. Loxandrus infimus in Texas   <Psyche,45,p.84.
38. The American Patrobini   <Ent.Amer.,18,135-187.

Davis, A. C.
34. Two n.sp. of Pleocoma   <Proc.Ent.Soc.Wash.,36,23-25.
34. Ins. in gr. squirrel burrows   <Bull.Br.Ent.Soc.,29,79-83.
34. A n.var. of Pleocoma   <Proc.Ent.Soc.Wash.,36,88-89.
35. Rev. of Pleocoma   <Bull.So.Cal.Ac.Sci.,33,123-130.
35. Rev. of Pleocoma   <Bull.So.Cal.Ac.Sci.,34,4-36.
35. (Myllocerus) Corigetus ? castaneus   <Bull.Br.Ent.Soc.,30,p.19.
35. Ins. in mushroon houses   <Proc.Ent.Soc.Wash.,36,1934
(1935),p.269.

Davis, J. J.
33. Insects of Indiana for 1932   <Proc.Ind.Ac.Sci.,42,213-225.
34. Insects of Indiana for 1933   <Proc.Ind.Ac.Sci.,43,195-201.
35. Insects of Indiana for 1934   <Proc.Ind.Ac.Sci.,44,198-206.
36. Insects of Indiana for 1935   <Proc.Ind.Ac.Sci.,45,257-268.
37. Insects of Indiana for 1936   <Proc.Ind.Ac.Sci.,46,230-239.

Dawson, R. W.
33. N.spp. of Serica, VII   <Jl.N.Y.Ent.Soc.,41,435-440.

Dean, Henry Lee
36. Rec. of dodder gall-weevel in Iowa   <Proc.Iowa Ac.Sci.,43,139-141.

Dejean, P. F. M. A. and Boisduval, J. B. A.
29. Iconogr. et hist nat. Col. Eur.   Paris, I, 1829, 400 pp.

DeLeon, D.
34. Annot. list of parasites...pine beetle <Can.Ent.,66,51-61.

Delkeskamp, K.
—. Cantharidæ   <Junk Col.Cat.(in press).

Denier, P. C. L.
34. Ensayo de class. de los Pyrota   <Rev.Soc.Ent.Argent.,6,49-75.
35. Col. Amer. fam. Meloidarum enum. syn.<Rev.Soc.Ent.Argent.,7,139-176.

Desbrochers des Loges, J.
72. Monog. Phyllobiides d'Europe...   <L'Abeille,11,659-748.
77. Monog. des Balanin. et Anthonom.d'Eur.<Ann.Soc.Ent.Fr.,(4),8,331-368,
411-470.
08. Faunule des Col. de la France...   <Le Frelon, 16, 1-36.
08. Desc. Curcul. nouv...d'Eur.   <Le Frelon, 16, 63-80.

Dobzhansky, T.
33. Geogr. var. in lady-beetles   <Amer.Nat.,67,97-126.
35. List of Coccinellidæ of B.C.   <Jl.N.Y.Ent.Soc.,43,331-336.

Dodge, H. R.
37. Occurr. of two Eur. Nitidulids in Wis. <Ent.News,48,p.285.
38. The bark beetles of Minnesota   <Univ.Minn.Ag.Exp.Sta.,Tech.
Bull.132,60 pp.
38. Coccidula suturalis synonymy   <Ent.News,49,221-222.

Dohrn, C. A.
78. Exotisches   <Stett.Ent.Zeit.,39,359-364.

d'Orchymont, A.
33. Contrib. á...Palpicornia. VIII.   <Bull.& Ann.Soc.Ent.Belg.,73,
271-313.
34. Notes sur quelques Helophores   <Bull.& Ann.Soc.Ent.Belg.,74,
251-261.
37. Quelques syn...Hydrophilidæ   <Bull.Mus.R.Hist.Nat.Belg.,12,
no.23,29 pp.

Downes, W.
38. ...Sitona lineatus in B.C.   <Can.Ent.,70,p.22.

Duges. E.
70. Desc. algunos Meloideos indigenos   <La Naturaleza,1,157-168.
89. Sinopsis de los Meloideos...**Mex.**   <An.Mus.Michoacano,2,34-40, 49-114.

Duncan, D. K.
34. Acmæodera papagonis   <Bull.Br.Ent.Soc.,29,p.64.
34. Random notes of Ariz. Collr.   <Bull.Br.Ent.Soc.,28,1933 (1934), 229-232.

DuVal, P. N. C. Jacquelin
54. Desc...esp. nouv. de Col.   <Bull.Soc.Ent.Fr.,1854,36-38.

Easterling, G. R.
34. Study of ins. fauna of Chio   <Chio Jl.Sci.,34,129-146.

Ebeling, W.
35. N. Scolytid in...lemon trees   <Pan-Pac.Ent.,11,21-23.

Eggers, H.
33. Zur synonymie der Borkenkäfer   <Ent.Nachrbl.,7,17-20.
34. Zur synonymie der Borkenkäfer   <Ent.Nachrbl.,8,25-29.

Eichhoff, W.
68. Neue Borkenkäfer   <Berl.Ent.Zeit.,1868,273-282.

Eidmann, H.
35. Zur...Insektenfauna von Südlabrador   <Arb.morph.& tax.Ent.B.-D., 2, 81-105.

Engelhardt, G. P.
36. Cymatodera californica...in Ariz.   <Bull.Br.Ent.Soc.,31,p.98.

Eppelsheim, E.
95. Beitrag zur Staphyl. West-Afrika   <Deutsche Ent.Zeit.,1895, 113-141.

Erichson, W. F.
36. Syst...der Borkenkäfer (Bostrychidæ)   <Arch.Naturg.,1,1835 (1836), 45-65.
42. Beitr.zur Fauna von Vandiemensland   <Arch.Naturg.,8,(2),83-287.

Essig, E. O.
33. Nomencl. of the vegetable weevil   <Science,77,605-606.

Everts, J. E.
22. Coleoptera Neerlandica   <'s-Gravenhage,1898-1922, vol.3.

Everly, R. T.
38. Spiders & insects...with sweet corn...<Ohio Jl.Sci.,38,136-148.

Exline, H. and Hatch M. H.
34. Note on food of black-widow spider   <Jl.N.Y.Ent.Soc.,42,449-450.

Fairmaire, L.
49. Essai...Coleop. de la Polynesie   <Rev.Mag.Zool.,(2).1, 277-291, 352-365. 410-422. 445-460, 504-516.
88. Desc. de Col. de l'Indo-Chine   <Ann.Soc.Ent.Fr.,(6),8,333-373.

Fairmaire, L. and Coquerel, C.
60. Essai...Coleop. de Barbarie   <Ann.Soc.Ent.Fr.,(3),8,145-176.

Fairmaire, L. and Germain, P.
61. Coleoptera Chilensia, II.   <Paris, 1860-1861, 85pp.

Falderman, F.
35. Col...China bor., Mongolia...   <Mem.Ac.St.Petersb.,2,337-464.

Fall, H. C.
19. Change of names   <Ent.News. 30. p.26.
31. Fabrician types...in Glasgow   <Ent.News.42.263-267.
33. Agonoderus pallipes   <Ent.News.44. 102-104.
33. New Coleoptera, XVI   <Can.Ent..65..229-234.
34. On...N.A. Elateridæ   <Jl.N.Y.Ent.Soc.,42,7-36.
34. Rev. of N.A. Agathidium   <Ent.Amer. 14.1933 (1934).99-131.
34. A new Trichochrous   <Can.Ent..66.142-143.
34. A new name and...notes   <Pan-Pac.Ent..10.171-174.
34. On two species of Ludius   <Bull.Br.Ent.Soc.,28,1933 (1934), 188-192.
34. New Buprestid from Fla. keys   <Ent.News.45.193-195.
36. On certain spp. of Cantharis   <Pan-Pac.Ent..12.179-183.
37. A new Agaporus   <Ent.News,48.10-12.
37. Misc. notes and desc.   <Can.Ent..69.29-32.
37. N.A. spp. of Nemadus   <Jl.N.Y.Ent.Soc.,45,335-340.

Fall, H. C. and **Davis, A. C.**
34. Col. of Santa Cruz Id., Cal.   <Can.Ent.,66,143-144.

Fattig, P. W.
33. Bomb. beetles of...Dicælus & Harpalus <Can.Ent.,65,190-191.
35. Coleoptera of Georgia   <Ent.News,46,153-160.
36. Coleoptera of Georgia   <Ent.News,47,15-20, 233-238.
37. Coleoptera of Georgia   <Ent.News,48,4-10, 250-255.
37. Bombard. beetles...Harpalus, Pasi- <Can.Ent.,69,47-48.
    machus

Fauvel, A.
78. Staphyl. des Moluques et N.Guinee   <Ann.Mus.Civ.Stor.Nat.Genova,
    12,171-315.
91. Voyage...Simon au Venez. Staphyl.   <Rev.d'ent.,10,87-127.

Felt, E. P.
37. Balloons as indicators of insect drift <Bartlett Tree Res.Lab.Bull.2.
38. Wind drift and dissem. of insects   <Can.Ent.,70,221-224.

Felt, E. P. and **Bromley, S. W.**
37. Xyleborus germanus in N. A.   <Bartlett Tree Res.Lab.Bull.2.

Felt, E. P. and **Chamberlain, K. F.**
35. Cccurrence of insects at some height <N.Y.St.Mus.Circ.17,1935, 70 pp.

Fenyes, A.
14. Sauter's Formosa-Ausb.: Aleocharinæ <Arch.Naturg.,80,A,2,45-55.

Ferris, G. F.
35. Prothoracic pleurites of Col.   <Ent.News,46,63-68, 93-95.

Fiedler, C.
35. Rüsslergattung Cœlasternus   <Ent.Nachrbl.,9,65-116, 117-148,
    157-173.

Fiori, A.
94. Alcune n.sp. & var. Staphyl...Italia <Nat.Sicil.,13,86-100.
15. App. fna. Col...Italia merid. e Sicilia <Riv.Col.Ital.,13,5-17.

Fisher, W. S.
34. New Anobiid from Canada   <Can.Ent.,66,275-276.
35. New genus of Buprestidæ from Utah <Proc.Ent.Soc.Wash.,37,117-118.
38. New Anobiid from Alaska   <Jl.Wash.Ac.Sci.,28,26-27.
38. New Anobiid injurious to books   <Proc.Ent.Soc.Wash.,40,43-44.

Fleischer, A.
08. Bestimm.—Tab...Liodini   <Verh.Nat.Ver.Brunn,46,1907
    (1908),3-63.
09. Notiz über einige Colon-Arten   <Wiener Ent.Zeit.,28, p.246.

Fleutiaux, E. and **Salle, A.**
89. List des Col. de Guadeloupe   <Ann.Soc.Ent.Fr.,(6),9,351-484.

Forbes, W. T. M.
22. Wing-venation of the Coleoptera   <Ann.Ent.Soc.Amer.,15,328-352.

Fox, H.
34. Distrib. of Jap. beetle in 1932 and 1933 <Jl.Econ.Ent.,27,461-473.

Friend, R. B.
29. Asiatic beetle in Conn.   <Conn.Agr.Exp.Sta.Bull.304.

Frost, C. A.
33. Hister semisculptus   ˙<Bull.Br.Ent.Soc.,28,p.159.
34. Trotting the bogs...   <Bull.Br.Ent.Soc.,29,1933 (1934),
    233-234.
35. Three beetles from Labrador   <Can.Ent.,67, p.19.
38. Boreaphilus americanus   <Bull.Br.Ent.Soc.,33,p.58.
38. Silpha americanus   <Bull.Br.Ent.Soc.,33,p.70.
38. Hoplia equina .   <Bull.Br.Ent.Soc.,33,p.107.

Frost, S. W.
16. Biol. notes on Ceutorhynchus marginatus<Jl.N.Y.Ent.Soc.,24,243-253.
24. Frogs as insect collectors   <Jl.N.Y.Ent.Soc.,32,173-185.
24. Leaf-mining habit in Col.   <Ann.Ent.Soc.Amer.,17,437-467.
29. Col. taken from bait traps   <Ann.Ent.Soc.Amer.,22,427-437.
36. Summary of ins...liquid baits   <Ent.News,47,64-68, 89-92.
37. New records from bait traps   <Ent.News,48,201-202.
38. Frederick Valentine Melsheimer   <Lancaster Hist.Soc.,41,(6), 1937
    (1938), 164-168.

Füessly, J. C.
75. Verz...Schweizerischen Insekten — Zurich,1775, 62 pp.

Gahan, C. J.
00. Stridulating organs in Col. — <Tr.Ent.Soc.Lond..1900,433-452.
10. Notes on Cleridæ... — <Ann.Mag.Nat.Hist.,(8),5,55-76.

Gaines, J. C.
33. Trap coll. of ins. in cotton in 1932 — <Bull.Br.Ent.Soc.,28,47-53.
33. Notes on Coccinellidæ... — <Jl.N.Y.Ent.Soc.,41,263-264.
34. Notes on Texas Coccinellidæ — <Bull.Br.Ent.Soc.,28,1933 (1934), 211-215.

Gardner, J. C. M.
35. Histeridæ: Niponiinæ — <Gen.Ins.,fasc.202, 6 pp.

Gautier des Cottes,—
62. Gen.n. Staphyl...de Syrie et d'Eur. — <Ann.Soc.Ent.Fr.,(4),2,75-78.

Gavoy, L.
26. Nouv. addit...au catal. de l'Aude — <Bull.Soc.Etud.sci.Aude,29,1925 (1926),143-163.

Gebien, H.
36. Katalog der Tenebrioniden. I.

Gehin, J. B.
85. Cat. syn. et syst...Carabides — <Remiremont,1885, 38 + 103 pp.

Geiser, S. W.
33. G. W. Belgrage's Texas localities — <Ent.News,44,127-132.

Gemminger, M.
70. [List of preoccupied names] — <Col.Hefte,6,119-124.

Gerhardt, J.
10. Neuheiten der Schles. Käferf...1909 — <Deutsche Ent.Zeit..1910,554-557.

Germar, E. F.
21. Genera q. Curculionitum... — <Mag.Ent.,4,291-345.

Gibson, A. and Crawford, H. G.
33. Norman Criddle — <Can.Ent.,65,193-200.

Gistel, J. N. F. X. (also Gistl)
48. Naturg. des Thierreichs... — Stuttgart,1848,216 pp.

Glasgow, R. D.
16. Phyllophaga: Rev. of Syn. — <Bull.Ill.Lab.Nat.Hist..11,365-379.

Glen, R.
35. Morph. of Ludius æripennis destructor — <Can.Ent.,67,231-238.

Glover, T.
78. Illustr. of N.A. Ent., 1878, Col.

Good, N. E.
36. Flour beetles of genus Tribolium — <U.S.Dpt.Agr.,Tech.Bull.498, 57pp.

Gory, H. L. and Percheron, A. R.
33. Monog. des Cetoines... — Paris, 1833, 403 pp.

Graham, S. A.
22. Study of wing venation of Col. — <Ann.Ent.Soc.Amer.,15,191-200.

Graves, H. W.
38. Hawaiian Elaterid...in Calif. — <Pan-Pac.Ent.,14,p.91.

Greiner, J.
37. Malachiidæ — <Junk Col.Cat.,pars 159.

Gridelli, E.
36. Tredic. contrib...Staphylinini — <Boll.Soc.Ent.Ital.,68,147-156.

Guignot, F.
28. Notes sur Haliplus du gr. fulvus — <Ann.Soc.Ent.Fr.,97,133-151.
35. Douziene note...Hydrocanthares — <Bull.Soc.Ent.Fr.,1935,36-40.

Günther, K. and Zumpt, F.
33. Curculionidæ: Tanymecinæ — <Junk Col.Cat.,pars 131.

Haag-Rutenberg,—
80: Beitr...Canthariden — <Deutsche Ent.Zeit..24,17-90.

Hamilton, J.
96. (Notes and descriptions) — <Leng, Lamiinæ of N.A.

Hardy, G. A.
36. Notes on Vanc.Id. Cerambycidæ   <Rpt.Prov.Mus.N.H.,B.C., 1935
                                          (1936),34-35.
Harold, F. E.
63. Note sur esp. mex. Phanæus   <Ann.Soc.Ent.Fr.,(4),3, 161-176.
Harris, T. W.
26. Minutes...hist...Melolontha...   <Mass.Agr.Repos.,10,1-12.
Hatch, M. H.
33. Species of Miscodera   <Pan-Pac.Ent.,9,7-8.
33. Records of Col. from Montana   <Can.Ent.,65,5-15.
33. Studies on Hydroporus   <Bull.Br.Ent.Soc.,28,21-27.
33. Notes on Carabidæ   <Pan-Pac.Ent.,9,117-121.
33. Stud...Leptodirinæ (Catopidæ)   <Jl.N.Ý.Ent.Soc,.41,187-239.
34. Ptinus tectus in America   <Bull.Br.Ent.Soc.,28,1933 (1934),
                                          200-202
35. Two...blind beetles from n.e.Ore.   <Pan-Pac.Ent.,11,115-118.
35. Monillipatrobus = Psydrus   <Pan-Pac.Ent.,11,118-119.
35. New sub-alpine Halticini   <Ent.News,46,276-278.
36. Studies on Leiodidæ   <Jl.N.Y.Ent.Soc.,44,33-41.
36. Studies on Pterostichus   <Ann.Ent.Soc.Amer.,29,701-706.
37. Two n.spp. of Helmidæ...Mont.   <Ent.News,49,16-19.
37. Note on Coleop. fauna of Alaska   <Pan-Pac.Ent.,13,p.63.
38. Rec. of Histeridæ from Iowa   <Jl.Kans.Ent.Soc.,11,17-20.
38. Anobium punctatum in Wash.   <Jl.Econ.Ent.,31,p.545.
38. Theophrastus as econ. entom.   <Jl.N.Y.Ent.Soc.,46,223-227.
38. N.sp.of Donacia from Wash.   <Pan-Pac.Ent.,14,110-112.
38. Rpt. on Col. coll...Aleutian Is.   <Pan-Pac.Ent.,14,145-149.
38. Biblio.cat...Arach. & Ins. of Wash.   <Univ.Wash.Pub.Biol.,1,167-223.
38. Two n.sp. of Helmidæ...Mont.   <Ent.News, 49,16-19.
38. Col. of Wash.: Cicindelinæ   <Univ.Wash.Pub.Biol.,1,229-238.
Hatch, M. H. and Beer, F. M.
38. N.sp. of Dicerca from Wash.   <Pan-Pac.Ent.,14,p.151.
Hatch, M. H. and Rueter, W.
34. Col. of Wash.; Silphidæ   <Univ.Wash.Pub.Biol.,1,147-162.
Haydon, S.
31. Dynastes hercules and tityrus   <Proc.Nat.Hist.Soc.Md.,August
                                          1931, 1-5.
Hayes, W. P.
35. Distrib. of Trichobaris insolita   <Bull.Br.Ent.Soc.,30,p.28.
36. Two n.sp. of Cosmobaris   <Jl.Kans.Ent.Soc.,9,26-29.
Hayes, W. P. and Kearns, C. W.
34. Pretarsus in Coleoptera   <Ann.Ent.Soc.Amer.,27,21-33.
Heikertinger, F. and Csiki, E.
—. Chrysomelidæ: Halticinæ   <Junk Col.Cat., in press.
Hendrickson, G. O.
30. Biol. notes on Microrhopala vittata   <Can.Ent.,62,98-99.
30. Stud. on ins. fna. of Iowa prairies   <Ia.St.Coll.Jl.Sci.,4,49-179.
31. ...insect fna. of Iowa prairies   <Ia.St.Coll.Jl.Sci.,5,195-209.
34. Mordellidæ of Iowa prairies   <Bull.Br.Ent.Soc.,28, 1933 ((1934),
                                          p.193.
Henriksen, K. L.
35. Insects and Acarina   <Meddel.om Groenl,.104,no.16,1-9.
Henshaw, S.
98. Ent. writings of G. H. Horn   <Tr.Amer.Ent.Soc.,25,1893-99,
                                          xxv-lxxii.
Hetschko, A.
33. Tretothoracidæ Jacobsoniidæ, Gnosti-<Junk Col.Cat.,pars 127 (part)
    dæ, Cavicoxumidæ
Heuer, A.
26. Eine var. von Archicarabus nemoralis   <Intern.Ent.Zs.,19,p.332.
Hincks, W. D. and Dibb, J. R.
35. Passalidæ   <Junk Col. Cat., pars 142.
Hinman, F. G. and Larson, A. O.
35. Ins. coll. in flight traps in Ore.   <Ent.News, 46,147-153.

84 BIBLIOGRAPHY

Hinton, H. E.
  30. Cbs. on two California beetles   <Pan-Pac.Ent.,7,94-95.
  34. N.spp. of Terapus from N.A.   <Ent.News,45,270-272.
  34. Helichus puncticollis in Ariz.   <Can.Ent.,66,p.72.
  34. N.n. for Atænius consors   <Can.Ent.,66,p.119.
  34. Aphodius...cadaverinus group   <Can.Ent.,66,218-220.
  34. N.n. for Aphodium smithi   <Ent.News,45,p.277.
  35. Rev. of N.A. Pseudister   <Can.Ent.,67,11-15.
  35. N.spp. of N.A. Helichus   <Pan-Pac.Ent.,11,67-71.
  35. Addit. to Histeridæ of L.Cal.   <Can.Ent.,67,78-82.
  34. N.spp. of N.A. Aphodius   <Stylops,3,188-192.
  35. N.g. & n.sp. of Neotrop. Colydiidæ   <Rev.de Ent.,5,202-215.
  35. Notes on the Dryopoidea   <Stylops,4,169-179.
  35. Desc. n. neotrop. Histeridæ...   <Ann.Mag.Nat.Hist.,(10),15,584-
                                              592.

  35. Notes on Amer. sp. of Colydodes   <Ent.Mo.Mag.,71,227-231.
  36. N.g. & n.spp. of Elminæ   <Ent.Mo.Mag.,72,1-5.
  36. Desc. of n.g. & n.spp. of Dryopidæ   <Tr.R.Ent.Soc.Lond.,85,415-434.
  36. Syn...notes on Dryopidæ   <Ent.Mo.Mag.,72,54-58.
  36. Misc. stud. in neotrop. Colydiidæ   <Rev.de Ent.,6,47-97.
  36. Notes on Amer. Colydiidæ   <Ent.News,47,185-187.
  36. Stud. in Mex. & C.A. Eupariini   <Univ.Cal.Pub.Ento.,6,273-276.
  36. Notes on Lobogestoria   <Ent.Mo.Mag.,72,128-129.
  37. Helichus immsi, sp.n. and notes   <Ann.Ent.Soc.Amer.,30,317-322.
  37. Notes on Braz. Potamophilinæ &   <Ent.Mo.Mag.,73,95-100.
      Elminæ
  37. Addit. to neotrop. Dryopidæ   <Arb.morph.tax.Ent.B.-D.,4,93-
                                              111.
  37. Desc. n. N.A. Atænius...   <Ann.Mag.Nat.Hist.,(10),20,177-
                                              196.
  38. N.spp. of neotrop. Aphodiinæ   <Rev.de Ent.,8,122-129.
  39.

Hinton, H. E. and Ancona, H.
  34. Fna.Col. en nidos de Hormigos   <An.Inst.Biol.Mex.,6,243-248.
  35. Fna.Col. en nidos de Hormigos   <An.Inst.Biol.Mex.,6,307-316.

Hochhuth, J. H.
  51. Beitr. zur. Rüsselkafer Russlands   <Bull.Moscou,24,I,3-102.
  72. Enum...Kiew und Volhynien...   <Bull.Moscou,44,II,85-117.   ·
      Käfer, II.
  72. Enum...Kiew und Volhynien...   <Bull.Moscou,45,II,195-234. 283-322.
      Käfer, III.

Hoffmann, A.
  28. Aberr. nouv. de Carabus problematicus <Misc.Ent.,31,p.12.

Hoffman, C. H.
  35. Biol. & tax. of Trichiotinus   <Ent.Amer.,15,133-214.

Holway, R. T.
  35. Prel. note on pretarsus...   <Psyche,42,1-24.

Hope, F. W.
  35. Char. & desc. of n.g. & n.spp. of Col. <Tr.Zool.Soc.Lond.,1,91-112.

Hopping, R.
  33. N. Buprestid from B. C....   <Pan-Pac.Ent.,9,84-88.
  33. N. Col. from w. Canada IV   <Can.Ent.,65,281-286.
  34. A new Neobellamira   <Can.Ent.,66,115-116.
  34. A change of name   <Pan-Pac.Ent.,10,p.174.
  35. N. Col. from w. Canada V   <Can.Ent.,67,8-9.
  35. Cbs. on nomencl. & tax. of Col.   <Proc.Ent.Soc.Br.Col.,31, 1934
                                              (1945),33-35.
  35. Rev. of Mycterus   <Pan-Pac.Ent.,11,75-78.
  36. Rev. of Macropogon   <Pan-Pac.Ent.,12,45-48.
  36. Note on Trigonurus   <Bull.Br.Ent.Soc.,31,p.65.
  37. Lepturini of N.A. II   <Can.Dpt.Mines & Res.Bull.85,42
                                              pp.
  37. N. Col. from w. Canada VI   <Can.Ent.,69, 89-91.

Hopping, R. and Hopping, G. R.
  34. Rev. of Cephaloon   <Pan-Pac.Ent.,10.64-70.

Horn, W.
  14. 50 neue Cicindelinæ   <Arch.Naturg.,79,1913 (1914), A, 11,1-33.
  35. Cicind. of Mex.,W.I. & U.S.   <Pan-Pac.Ent.,11,65-66.
  38, 2000 Zeichnungen von Cicindelinæ   <Ent.Beihefte B.-D.,5,1-71.

[Howard, L. O.]
  35. E. A. Schwarz   <Dict.Amer.Biogr.,1935.

Hustache, A.
  22. Syn. & disp. de Pantomorus godmani <Bull.Soc.Ent.Fr.,1922,100-101.
  33. Deux n. Curculionides depredateurs   <Bull.Mus.Nac.Nat.Paris,5, 376-380.
  34. Curculionidæ: Zygopinæ   <Junk Col.Cat.,pars 136.
  36. Curculionidæ: Cryptorrhynchinæ   <Junk Col.Cat.,pars 151.
  38. Curculionidæ: Barinæ   <Junk Col.Cat.,pars 163.

Ingram, W. M.
  34. Field notes on...Cicindela...Cal.   <Jl.Ent. & Zool.,26,51-52.

Isely, D.
  20.   <U.S.D.A.Bull.901, 21 pp.

Jeannel, R.
  33. Trois Adelops nouv.de N.A.   <Bull.Soc.Ent.Fr.,38,251-253.
  37. Les Bembidiides endoges   <Rev.Franc.Ent.,4,1-23.
  37. Notes sur les Carabiques   <Rev.Franc.Ent.,3,241-399.

Jellison, W. J. and Philip, C. B.
  33. Faunæ of nests of magpie & crow   <Can.Ent.,65,26-31.

Jimenez, L. M.
  66. Dictamen...ins. pres. por Sr. Barranco<Gac.Med.Mex.,2,225-230.

Jones, M. P.
  35. Peculiar ins. situation on seashore   <Proc.Ent.Soc.Wash.,37,150-151.

Jordan, K.
  04. American Anthribidæ   <Nov.Zool.,11,242-309.

Joseph, G.
  68. Lathrobium (Centrocnemis) krniense n.sp.   <Berl.Ent.Zeit.,12,365-366.

Jules,—.
  35. Deleaster dichrous in N. A.   <Le Nat.Canad.,62,p.5.

Junk, W. (Publisher) and Schenkling, S. (Editor)
  Coleopterorum Catalogus.
  125. Curculionidæ: Pissodinæ. K. W. von DallaTorre, S. Schenkling, and G. A. K. Marshal, 1932, 29 pp.
  126. Carabidæ: Harpalinæ VIII (Conclusion). E. Csiki, 1933, pp. 1599-1933.
  127. Carabidæ: Carabinæ III (Corrigenda et Addenda). E. Csiki, 1933, pp. 623-648.
    Gnostidæ. A. Hetschko, 1933, 1 p.
    Tretothoracidæ. A. Hetschko, 1933, 1 p.
    Jacobsoniidæ. A. Hetschko, 1933, 1 p.
    Cavicoxumidæ. A. Hetschko, 1933, 1 p.
  128. Lycidæ. R. Kleine, 1933, 145 pp.
  129. Staphylinidæ VII. O. Scheerpeltz, 1933, pp. 989-1500.
  130. Staphylinidæ VIII. O. Scheerpeltz, 1934, pp. 1501-1881.
  131. Curculionidæ: Tanymecinæ. K. Günther and F. Zumpt, 1933, 131 pp.
  132. Buprestidæ III. J. Obenberger, 1934, pp. 571-781.
  133. Telegeusidæ. S. Schenkling, 1934, 1 p.
    Biphyllidæ. S. Schenkling, 1934, 7 pp.
    Aculognathidæ. S. Schenkling, 1934, 1 p.
    Hemipeplidæ. S. Schenkling, 1934, 1 p.
    Scalidiidæ. S. Schenkling, 1934, 1 p.
  134. Curculionidæ: Cleoninæ. E. Csiki, 1934, 152 pp.
  135. Curculionidæ: Gymnetrinæ, Nanophyinæ. A. Klima, 1934, 68 + 26 pp.
  136. Curculionidæ: Zygopinæ. A. Hustache, 1934, 96 pp.
  137. Curculionidæ: Hyperinæ. E. Csiki, 1934, 66 pp.
  138. Curculionidæ: Cioninæ, Tychiinæ. A. Klima, 1934, 21 + 61 pp.
  139. Curculionidæ: Anthonominæ, Læmosaccinæ. S. Schenkling, and G. A. K. Marshall, 1934. 82 + 8 pp.
  140. Curculionidæ: Erirrhininæ. A. Klima, 1934, 167 pp.

141. Ectrephidæ. S. Schenkling, 1935, 4 pp.
Curculionidæ: Magdalinæ. S. Schenkling, 1935, 31 pp.
142. Passalidæ. W. D. Hincks and J. R. Dibb, 1935, 118 pp.
143. Buprestidæ IV. J. Obenberger, 1935, pp. 785-934.
144. Curculionidæ: Otidocephalinæ, Ithycerinæ, Belinæ, Petalochilinæ, Oxycoryninæ. K. W. von DallaTorre and E. Voss, 1935, 13 + 2 + 14 + 2 + 2 pp.
145. Curculionidæ: Alophinæ, Diabathrarinæ, Rhynchæninæ, Ceratopinæ, Tri-
gonocolinæ, Xiphaspidinæ, Nerthopinæ, Euderinæ, Camarotinæ, Acicnemidinæ. A. Klima, 1935, 14 + 4 + 36 + 3 + 3 + 1 + 2 + 1 + 2 + 10 pp.
146. Curculionidæ: Cholinæ, Tachygoninæ, Antliarrhininæ, Ulomascinæ, Epi-
pedinæ, Pyropinæ. A. Klima, 1936, 32 + 4 + 1 + 1 + 1 pp.
147. Curculionidæ: Brachyderinæ I. K. W. von DallaTorre and M. & F. van
Emden, 1936, 132 pp.
148. Curculionidæ: Otiorrhynchinæ I. C. Lona, 1936, 226 pp.
149. Curculionidæ: Rhynchophorinæ, Cossoninæ. E. Csiki, 1936, 199 pp.
150. Curculionidæ: Prionomerinæ, Aterpinæ, Amalactinæ, Haplonychinæ, Omophorinæ. S. Schenkling and G. A. K. Marshall, 1936, 11 + 9 + 3 + 8
+ 2 pp.
151. Curculionidæ: Cryptorrhynchinæ. A. Hustache, 1936, 317 pp.
152. Buprestidæ V. J. Obenberger, 1936, pp. 935-1246.
153. Curculionidæ: Brachyderinæ II. K. W. von DallaTorre and M. &. F. van
Emden, 1937, 133-196.
154. Curculionidæ: Rhadinosominæ, Trachodinæ, Raymondionyminæ. S.
Schenkling and G. A. K. Marshall, 1937, 2 + 4 + 6 pp.
155. Dàsytidæ: Dasytinæ. M. Pic, 1937, 130 pp.
156. Scarabæidæ: Dynastinæ. G. J. Arrow, 1937, 124 pp.
157. Buprestidæ VI. J. Obenberger, 1937, pp. 1249-1714.
158. Curculionidæ: Mesoptiliinæ, Rhynchitinæ I. K. W. von DallaTorre and
E. Voss, 1937, 56 pp.
159. Malachiidæ. J. Greiner, 1937, 199 pp.
160. Curculionidæ: Otiorrhynchinæ II. C. Lona, 1937, pp. 229-412.
161. Bostrychidæ. P. Lesne, 1938, 84 pp.
162. Curculionidæ: Otiorrhynchinæ III. C. Lona, 1938, pp. 415-600.
163. Curculionidæ: Barinæ. A. Hustache, 1938, 219 pp.
164. Curculionidæ: Brachyderinæ III. M. & F. van Emden, in press.
——. Cantharidæ. K. Delkeskamp, in press.
——. Chrysomelidæ: Halticinæ. F. Heikertinger and E. Csiki, in press.
——. Curculionidæ: Rhynchitinæ II. E. Voss, in press.

Kaston, B. J.
36. Morph. of...Hylurgopinus rufipes   <Conn.Agr.Exp.Sta.Bull.387,
613-650.

Kato, M.
35. On...Curculio dentipes...larval state   <Sci.Rpts.Tohoku Univ.,(4),10,
515-553.

Keifer, H. H.
31. Misc. insect notes   <Mo.Bull.Cal.Dpt.Agr.,20,(7),
470-472.
33. Pac. Cst. Ctiorhynchid larvæ   <Ent.Amer.,13 1932(1933),45-85.

Kirby, W.
25. Desc. of some ins...MacLeay's doc-<Tr.Linn.Soc.Lond.,14,93-110.
trine...

Kleine, R.
33. Lycidæ   <Junk Col.Cat.,pars 128.
37. Best.-Tab. der Brenthidæ   <Ent.Nachrbl.,11,17-29.

Klima, A.
34. Curculionidæ: Gymnetrinæ, Nano-   <Junk Col.Cat.,pars 135.
phyinæ
34. Curculionidæ: Cloninæ, Tychinæ   <Junk Col.Cat.,pars 138.
34. Curculionidæ: Erirrhininæ   <Junk Col.Cat.,pars 140.
35. Curculionidæ: Alophinæ, Diabathrarli-
næ, Rhynchæninæ, Ceratopinæ, Trigo-
nocolinæ, Xiphaspidinæ, Nerthopinæ,
Euderinæ, Camarotinæ, Acicnemidinæ.<Junk Col.Cat.,pars 145.
36. Curculionidæ: Cholinæ, Tachygoninæ,
Antliarrhininæ, Ulomascinæ, Epipedi-
næ, Pyropinæ.   <Junk Col.Cat., pars 146.

Klug, J. C. F.
  42. Versuch einer System...der Clerii <Abh.k.Akad.Wiss.Berl., 1840-42, 259-397.

Knight, H. H.
  36. Rec. of so. ins. moving northward <Ann.Ent.Soc.Amer.,29,578-580.

Knowlton, G. F.
  33. Ladybird beetles as predators... <Can.Ent.,65,241-243.
  34. Distrib. notes on Utah Coleoptera <Jl.Kans.Ent.Soc.,7,79-86.
  36. Distrib. notes on Utah Col. II. <Jl.Kans.Ent.Soc.,9,107-111

Knowlton, G. F. and Smith, C. F.
  35. Notes on Utah Scarab.& Chrysom. <Ent.News,46,241-244.

Knowlton, G. F. and Thatcher, T. O.
  36. Notes on wood-boring insects <Proc.Utah Ac.Sci.A. & L.,13, 277-281.

Knowlton, G. F. and Thomas, W. L.
  34. Some Cache Valley Utah insects <Proc.Utah Ac.Sci.A. & L.,11, 245-246.

Knull, J. N.
  34. Five n.sp. of Coleoptera <Ent.News,45,9-13.
  34. Two new Arizona Coleoptera <Ent.News,45,68-70.
  34. Notes on Coleoptera, no.4 <Ent.News,45,207-212.
  34. New Coleoptera <Chio Jl.Sci.,34,333-336.
  35. Four new Texas Coleoptera <Ent.News,46,96-99.
  35. New Coleoptera <Ent.News,46,189-193.
  36. Five new southwestern Coleoptera <Ent.News,47,73-75, 105-108.
  37. Notes on Col. with desc. of n.spp. <Ent.News,48,15-17, 36-42.
  37. New s.w. Buprestidæ & Cerambycidæ <Chio Jl.Sci.,37,301-309.
  37. N.spp. of Paratyndaris... <Ann.Ent.Soc.Amer.,30,252-257.
  38. Five new species of Coleoptera <Chio Jl.Sci.,38,97-100.
  38. Four new Coleoptera <Ent.News,49,19-22.
  38. New s.w. Buprestidæ & Cerambycidæ <Ann.Ent.Soc.Amer.,31,135-143.
  38. A new Acmæodera <Ent.News,49,p.228.

Koch, C.
  38. Wiss.Ergeb...Exp.nach Aegypten... <Pubb.Mus.Ent.Pietro-Rossi,1, 115-232.

Kraatz, G.
  51. Verz...Er. Käfer Mark Brand... <Stett.Ent.Zeit.,12,283-286, 291-295.
  57. Naturg. der Insect. Deutschl., 1857 (not 1858).
  60. Ueber die Gattung Diochus Er. <Wien.Ent.Monatschr.,4,25-28.
  78. [Notes on Carabus] <Deutsche Ent.Zeit.,22,p.158.
  78. Ueber Orinocarabus... <Deutsche Ent.Zeit.,22.327-335.
  79. Ueber Bockkäfer Cst-Sibiriens... <Deutsche Ent.Zeit.,23,77-117.

Krauss, N. L. H.
  37. Study of Glyptoscelis in N.A. <Univ.Cal.Pub.Ento.,7,21-32.

Krynicky, J.
  32. Enum. Coleop. Rossiæ merid... <Bull.Moscou,5,69-179.

Kuntzen, H.
  33. Aus der Verbreitungstats...Scarabæi-<Mitt.Zool.Mus.Berl.,19,458-474.
  den...
  37. Ueber Arrhenodes minutus... <Mitt.Zool.Mus.Berl.,22,190-197

Laboissiere, V.
  29. Sur la subf. des Chlamydinæ <Bull.Soc.Ent.Fr.,1929,256-258.

Lameere, A.
  38. Evolution des Coléoptères <Bull.& Ann.Soc.Ent.Belg.,78, 355-363.

Landis, B. J. and Davidson, R. H.
  34. Prothetely in Epilachna corrupta <Ohio Jl.Sci.,34,147-149.

Landis, B. J. and Mason, H. C.
  38. Var. elytral marks of Epilachna vari-<Ent.News,49,181-184.
  vestis

Lane, M. C.
  38. A n.sp. of Eanus <Pan.-Pac.Ent.,14,188-191.

Lange, W. H.
  37. Annot. list of ins. of Jeffrey pine <Pan-Pac.Ent.,13,172-175.

Lapouge, G. V. de
  05. ...Carabes et Calosomes...Mongolie   <Bull.Mus.Hist.Nat.Paris,11, 301-306.
  08. Tableaux de determ...Carabus   <L'Echange, Rev.Linn.,24,18-21.
  24. Calosomes nouveaux...   <Misc.Ent.,28,37-44,.
  24. Carabes nouveaux...   <Misc.Ent.,28,145-192.
  32. Carabidæ: Carabinæ   <Gen.Ins.,fasc.192c.

LaRivers, I.
  38. Cysteodemus in s. Nevada   <Pan-Pac.Ent.,14,124-128.

Latham, R.
  34. Xyloryctes satyrus on Long Id.   <Bull.Br.Ent.Soc.,28,1933 (1934), p. 202.

Latreille, P. A.
  34. Distrib. method...Serricornes   <Ann.Soc.Ent.Fr.,3,113-170.

Lawson, P. B.
  35. A beetle new to Kansas   <Jl.Kans.Ent.Soc,8,p.26.

Leach, E. R.
  33. Two old and two new Pleocomas   <Pan-Pac.Ent.,9,184-187.
  36. Phileurus illatus Lec.   <Pan-Pac.Ent.,11,1935 (1936), p. 169.

Leconte, J. L.
  48. Fragmenta Entomologica   <Jl.Ac.N.S.Phila.,(2),1,71-93.
  78. Coleoptera of Michigan   <Pp. 593-669.

Leech, H. B.
  34. Almost a cannibal   <Bull.Br.Ent.Soc.,29,p.41
  35. Trichocera garretti and...predator   <Can.Ent.,67,182-183.
  35. B.C. rec. of Carabidæ & Hydrophilidæ   <Pan-Pac.Ent.,11,120-124.
  36. A rare Aphodius   <Bull.Br.Ent.Soc.,31,p.56.
  37. New N.A. Agabus, with notes   <Can.Ent.,69,146-150.
  37. ...names in vespilloides gr. of Nicro-<Bull.Br.Ent.Soc.,32,156-159. phorus
  38. Hibernation of Plectrura   <Pan-Pac.Ent.,14,p.68.
  38. N.sp. of Gyrinus; Dineutes robertsi   <Can.Ent.,70,59-61.
  38. N.sp. of Cœlambus from Calif.   <Pan-Pac.Ent.,14,84-86.
  38. Desc. 3 n.sp. Agabus from Hudson Bay<Can.Ent.,70,123-127.

Lefebvre, E.
  38. Nouvelles diverses   <Ann.Soc.Ent.Fr.,7,Bull.x-xili.

Leng, C. W.
  23. N.sp. and syn. of Statira   <Jl.N.Y.Ent.Soc.,31,184-188.

Leng, C. W. and Mutchler, A. J.
  33. 2nd & 3rd Suppl. to Leng Cat. of Col.   Mt.Vernon,N.Y.,1933, 112 p.
  34. Saprinus dimidiatipennis   · <Jl.N.Y.Ent.Soc.,42,p.86.

Lepeletier, A. L. M. and Serville, A.
  25. See Clivier

Lesne, P.
  35. Quelques precis. sur Hendecatomus   <Bull.Soc.Ent.Fr.,40,197-199.
  37. Notes rect. et syn...Bostrychids   <Bull.Soc.Ent.Fr.,42,238-240.
  38. Bostrychidæ   <Junk Col.Cat.,pars 161.

Letzner, K.
  50. Syst. Beschr. der Laubkäfer Schlesiens<Zs.Ent.Breslau,1847-50 (1850), 1-112.

Lewis, G.
  92. On the Japanese Cleridæ   <Ann.Mag.Nat.Hist.,(6),10,183-192.
  95. Lamellicorn Coleop. of Japan   <Ann.Mag.Nat.Hist.,(6),16,374-408.

Liebke, M.
  33. Amer. Arten der Gattung Zuphium   <Rev. de Ent.,3.461-472.
  34. Arten der Gattung Pseudaptinus   <Rev.de Ent.,4,372-388.
  36. Die Gattung Lachnophorus   <Rev.de Ent.,6.461-468.
  37. Denkschr. uber Carab.-Tr. Colliurini   <Festschr.f.Emb.Strand,4,37-141.

Lindahl, J. C.
  35. Acmæodera hepburni v. latiflava   <Pan-Pac.Ent.,11,p.61.

Linell, M. L.
  98. N.sp. of...Chrysom...Chlamydini   <Proc.U.S.Nat.Mus.,20,473-485.

Linell, M. L. and Schwarz, E. A.
  98. Coleoptera, in Stejneger's Asiatic Fur-Seal islands and fur-seal industry. pt.4, app.1.
Linsley, E. G.
  33. Two interesting new records   <Pan-Pac.Ent.,9,p.92.
  33. Obs. on swarming of Melanophila   <Pan-Pac.Ent.,9,p.138.
  33. A new Calif. Clerid beetle   <Pan-Pac.Ent.,9,p.95.
  33. Life hist. & habits of...Aulicus   <Biologist,15,(1),87-90.
  33. A n.sp. of Neoclytus from White Fir.   <Pan-Pac.Ent.,9,93-94.
  33. A n.sp. of Monochamus from Calif.   <Can.Ent.,65,118-119.
  33. Eur. Longicorn new to Cal.   <Pan-Pac.Ent.,9, p.170.
  34. Rev. of Atimia with...n.sp.   <Pan-Pac.Ent.,10,23-26.
  34. Stud. in Cerambycidæ of L.Cal.   <Pan-Pac.Ent.,10,59-63.
  34. Rev. of Pogonocherini of N.A.   <Ann.Ent.Soc.Amer.,28,73-103.
  34. Notes & desc. of w.Amer.Cerambycidæ <Ent.News,45,161-165, 181-185.
  34. New longicorn...Lamiinæ   <Bull.Br.Ent.Soc.,28,1933 (1934), 183-185.
  35. Stud. in Longicornia of Mex.   <Tr.Amer.Ent.Soc.,61,67-102.
  35. N.spp. of Pleocoma...   <Pan-Pac.Ent.,11,11-15.
  35. Occurr. of some Cal. Cerambycidæ   <Pan-Pac.Ent.,11,p.15.
  35. Dectes spinosus   <Pan-Pac.Ent.,12,p.74.
  35. Notes & desc. of w.Amer. Ceramb.II.   <Ent.News,46,161-166.
  36. Note on...Hesperorhipis albofasciatus   <Pan-Pac.Ent.,12,p.110.
  36. Obs. on habits of...longicorns   <Pan-Pac.Ent.,12,199-200.
  36. Hibernation of Cerambycidæ   <Pan-Pac.Ent.,12,p.119.
  36. Studies in Aulicus   <Univ.Cal.Pub.Ento.,6,249-262.
  36. Prel. stud. of N.A. Phoracanthini &<Ann.Ent.Soc.Amer.,29,461-479. Sphærionini
  37. Notes & desc. of w.Amer. Ceramb.III.   <Ent.News,48,63-69.
  38. Notes on...spp. of Pleocoma   <Pan-Pac.Ent.,14,49-58, 97-104.
  38. Syn. notes on N. A. Cerambycidæ   <Pan-Pac.Ent.,14,105-109.
  38. Longevity in Cerambycidæ   <Pan.-Pac.Ent.,14,p.177.
  38. Ins. types from L.Cal. in Cal.Ac.Sci.   <Pan-Pac.Ent.,14,p.104.
Linsley, E. G. and Martin, J. O.
  33. Notes on longicorns from subtrop. Tex.<Ent.News,44,178-183.
Linsley, E. G. and Usinger, R. L.
  34. Insect collecting in Calif.   <Pan-Pac.Ent.,10,102-106.
  36. Insect collecting in Calif.II.   <Pan-Pac.Ent.,12,49-55.
Löben Sels, E. von
  34. Some obs. on Phalacrus politus...   <Jl.N.Y.Ent.Soc.,42,319-327.
Löding, H. P.
  33. Ala. Coleop. not generally listed...   <Bull.Br.Ent.Soc.,28,139-151.
  34. Dorcus brevis in Alabama   <Bull.Br.Ent.Soc.,29,p.36.
  34. New & rare beetles on sand beach   <Bull.Br.Ent.Soc.,29,p.43.
  34. Turpentine orchards...coll. Coleop.   <Bull.Br.Ent.Soc.,29,p.98.
  35. Geotrupes ulkei   <Bull.Br.Ent.Soc.,30,p.108.
Lona, C.
  36. Curculionidæ: Otiorrhynchinæ I.   <Junk Col.Cat.,pars 148.
  37. Curculionidæ: Otiorrhynchinæ II.   <Junk Col.Cat.,pars 160.
  38. Curculionidæ: Otiorrhynchinæ III.   <Junk Col.Cat.,pars 162.
Lucas, R.
  20. Cat. alphab. gen. et subg. Coleop...   Berlin, pars 1, 1920 696 pp.
Lynch-Arribalzaga, F.
  (See Arribalzaga in original catalog.)
Mader, L.
  38. Uber neue...Erotyliden   <Ent.Blatter,23,14-19.
Maklin, F. H.
  63. Mex. Arten...Statira   <Acta Soc.Sc.Fenn.,7,585-594.
Mank, E. W.
  34. Col. of Glacier Park, Mont.   <Can.Ent, 66,73-81.
  34. New species of Orobanus   <Pan-Pac.Ent, 10.121-124.
  37. Note on two Amer. Xylitas   <Can.Ent, 69,18-19.
  38. Rev. of Zilora   <Psyche,45,101-104.
Mann, W. M.
  21. Three new myrmec. Col.   <Proc.U.S.Nat.Mus.,59,547-552.

Mannerheim, C. G.
  30. Precis...Brachelytres   (Sometimes sep. paged 1-87.)
  37. Enum. des Buprestides...   <Bull.Moscou,10,(8),3-126.
Mansour, K.
  34. Phylogenetic classif. of Col.   <Bull.Soc.R.Ent.Egypte,17,1933
Marseul, S. A.   (1934),190-203.
  62. Essai monog...des Histerides   <Ann.Soc.Ent.Fr.,(4),2,5-48,
Marshall, M. Y.   437-516, 669-720.
  37. New Melyrid of genus Tanaops   <Bull.So.Cal.Ac.Sci.,35,164-165.
Maulik, S.
  36. Chrysomelidæ: Galerucinæ   <Fauna of Br.India, 1936, 648 pp.
Maydell, G. G.
  34. N.spp. of N.A. Meloidæ   <Tr.Amer.Ent.Soc.,60,327-336.
  35. N.sp. of blister beetle from Ariz.   <Jl.Wash.Ac.Sci.,25,p.72.
McClure, H. E.
  33. The Click-beetle's click   <Ent.News,44,145-147.
McKenzie, H. L.
  36. Anat. & syst. study of Anatis...   <Univ.Cal.Pub.Ento.,6,263-272.
Mead, A. R.
  38. N.ssp. of Donacia with key   <Pan-Pac.Ent.,14,113-120.
Meier, W.
  99. Ueber Abänder. einiger Col...   <Ent.Nachr.,25,97-102.
Meixner, J.
  35. Coleoptera, Strepsiptera   <Kukenthal's Handb. der Zool.,
Melsheimer, F. E.   4,(2,1),pp.1037-1382.
  44. (See 1846 in original bibliography.)
  46. (See 1847 in original bibliography.)
Ménétries, E.
  32. Cat. rais...voyage au Caucase et Perse St. Petersbourg, 1832.
  48. Desc. des ins. rec. par feu M. Lehman<Mem.Acad.St.Petersb.,6,1-112.
  06. (See Zaitzev—1906.)
Mequignon, A.
  34. Les Chelonarium d'Amerique...   <Arb.morph.tax.Ent.B.-D.,1,
         294-300.
  34. Les Chelonarium de l'Amer. contin... <Ann.Soc.Ent.Fr.,103,199-256.
  37. Syn. proposees par Embrik Strand <Bull.Soc.Ent.Fr.,42,276-279.
Meserve, F. G.
  35. Sphærites glabratus from B.C.   <Jl.Econ.Ent.,28,p.420.
  35. Necrophorus hybridus in Colo.   <Jl.Econ.Ent.,28,p.112.
  35. Compar. of markings in Cicindela... <Bull.Br.Ent.Soc.,30,p.162.
  36. Silphidæ of Nebraska   <Ent.News,47,132-134.
  36. Cicindelidæ of Nebraska   <Ent.News,47,270-275.
Metcalf, C. L.
  33. Thylodrias contractus in Ill.   <Jl.Econ.Ent.,26,509-510.
Milne, L. J.
  33. Notes on Pseudolucanus placidus   <Can.Ent.,65,106-114.
Moennich, H. C.
  34. Trapping for Coleoptera   <Bull.Br.Ent.Soc.,29,97-98.
  37. Col. found on Pleurotus fungi   <Bull.Br.Ent.Soc.,32,169-170.
Moore, I.
  37. List of beetles of San Diego ·Co.,Cal. <Occ.Pap.San Diego Soc.Nat.
         Hist.,2, 109 pp.
Moore, S.
  33. Calcsoma escaping by diving   <Bull.Br.Ent.Soc.,28,p.36.
Motschoulsky, V.
  45. Obs. sur musee ent. ...de Moscou <Bull.Moscou,1845,IV,332-388.
  46. Ins. de la Siberie...1839 et 1840   <Mem.Acad.Imp.Scl.St.Petersb.,
         Mem.Sav.Etr.,5,1-274.
  49. Col...M. Handschuh...Espagne... <Bull.Moscou,22,52-163.
  50. Die Käfer Russlands   Moscou, 1850.
  70. Enum. nouv. esp. de Col...VII   <Bull.Moscou,42,252-257.
  75. Enum. nouv. esp. de Col...XV   <Bull.Moscou,49,(1),139-155.

Mulsant, E
56. Hist. nat. Col. Fr., Heteromeres   <Ann.Soc.Linn.Lyon,(2),3,198-522.
70. Hist. Nat. Col. Fr., Lamellicornes   <Ann.Soc.Agr.Lyon,(4),2,241-650.
70. Hist. Nat. Col. Fr., Lamellicornes   <Ann.Soc.Agr.Lyon,(4),3,155-480.

Musgrave, P. N.
33. New species of Helmidæ   <Proc.Ent.Soc.Wash.,35,54-57.
35 Two n. Elmidæ from Puerto Rico   <Proc.Ent.Soc.Wash.,37,32-35.
35. Notes on collecting Dryopidæ   <Can.Ent.,67,61-64.
35. Syn. of Helichus of N.A.   <Proc.Ent.Soc.Wash.,37,137-145.

Netolitzky, F.
14. Die Bembidium in Winkler's Cat.   <Ent.Blatter,10,50-55,164-176.

Nicolay, A. S.
34. Answer to...E. G. Smyth...   <Ent.News,45,127-131,153-155.

Nicolay, A. S. and Weiss, H. B.
34. Notes on Carabidæ, incl. syn. of <Jl.N.Y.Ent.Soc.,42,193-212.
genera...

Nördlinger, H.
55. Die kleinen Feinde der Landwirthschaft .Stuttgart,1855,636 pp.

Nonfried, A. F.
90. Coleopterorum Species novæ   <Wien.Ent.Zeit.,9,76-78.

Nunenmacher, F. W.
34. Stud. among Coccinellidæ, 6.   <Pan-Pac.Ent.,10,17-21.
34. Stud. among Coccinellidæ, 7.   <Pan-Pac.Ent.,10,113-114.
37. Stud. among Coccinellidæ, 8.   <Pan-Pac.Ent.,13,182-183.

Obenberger, J.
34. Monog. du genre Taphrocerus   <Sbornik ent.Nar.Mus.Praze,
   12,5-62.
34. Buprestidæ III   <Junk Col.Cat.,pars 132.
35. Buprestidæ IV   <Junk Col.Cat.,pars 143.
36. Buprestidæ V   <Junk Col.Cat.,pars 152.
36. Buprestidæ VI   <Junk Col.Cat.,pars 157.
36. Festarbeit...Embrik Strand   <Festschr.E.Strand,1,97-145.
36. Synonymia Agrilorum III   <Casopis Cs.Spol.Ent.,33,p.139.

Ohaus, F.
12. Beitr. z. Kennt. Ruteliden X.   <Stett.Ent.Zeit.,73,273-319.
15. XVI Beitr. z. Kennt. Ruteliden.   <Deutsche Ent.Zeit.,1915,256-260.
33. Scarab.: Euchirinæ-Phænomerinæ   <Gen.Ins.,fasc.195.
34. Scarab.: Rutelinæ I   <Gen.Ins.,fasc.199a.

Olliff, A. S.
87. Rev. of Staphylinidæ of Australia III   <Proc.Linn.Soc.N.S.Wales,(2),2,
   471-512.

Pandellé, L.
69. Monog...Eur...Tachyporini   <Ann.Soc.Ent.Fr.,(4),9,261-366.

Park, O.
33. Food...of Tmesiphorus costalis   <Ent.News,44,149-151.
33. Ecol. study of...Limulodes   <Ann.Ent.Soc.Amer.,26,255-261.
35. Further rec. of beetles with ants   <Ent.News,46,212-215.
35. Beetles assoc. with Formica ulkei   <Psyche,42,216-231.

Parkin, E. A.
33. Larvæ of wood-boring Anobiidæ   <Bull.Ent.Res.,24,33-68.

Parker, H. L. and Smith, H. D.
34. Further notes on Eoxenos...   <Ann.Ent.Soc.Amer.,27,468-479.

Parsons, C. T.
36. Notes on N.A. Nitidulidæ: Pocadius   <Psyche,43,114-118.
38. Notes on N.A. Nitidulidæ: II.Cryptarcha<Psyche,45,96-100.
39. Notes on N.A. Nitidulidæ III:.   <Psyche,45,1938 (1939),156-164.

Pascoe, F. P.
62. Notices of n.g. & n.sp. of Col.   <Jl.Ent.,I,36-64.
63. Notices of n.g. & n.sp. of Col.   <Jl.Ent.,II,26-56.

Paykull, G. von
92. Monog. Curculionum Sueciæ   Upsala, 1792, 151 pp.

Pechuman, L. L.
37. Annot list of insects in (elm)   <Bull.Br.Ent.Soc.,32,8-21.
38. Prel. study of biol. of Scolytus sulcatus<Jl.Econ.Ent.,31,537-543.

Penafiel, A. and Barranco, —.
    66. 'Estudio...cantariadas Mexicanas I.   <Gac.Med.Mex.,2,225-227.

Percheron, A. R.
    35. Monographie des Passalus...   Paris, 1835, 107 pp.

Peringuey, L.
    08. Desr. Cat. S.Afr. Col. (Lucan.& Scarab.)<Tr.S.Afr.Phil.Soc.,13,1904-08.
                     (1908), 547-752.

Peyerimhoff, P. de
    Larves des Coleopteres...   <Ann.Soc.Ent.Fr.,102,77-106.

Peyron, E
    58. Cat. col. env. Tarsous...   <Ann.Soc.Ent.Fr.,(3),6,353-434.

Phillips, E. F.
    33. Ins. coll. on apple blossoms in N.Y.   <Jl.Agr.Res.,46,851-862.

Pic, M.
    19. Col. exotiques en partie nouveaux   <L'Echange,35,no.391,2-3,etc.
    33. Nouv. Coleopteres Amer.   <Bull.Soc.Ent.Fr.,38,292-294.
    37. Dasytidæ: Dasytinæ   <Junk Col.Cat.,pars 155.

Poll, M.
    33. Note sur la classif. des Coleop.   <Bull.& Ann.Soc.Ent.Belg.,73,
                     57-67.

Portevin, G.
    14. Rev. Silphides, Liodides, Clambides du<Ann.Soc.Ent.Belg.,58,212-236.
    Japon

Potter, C.
    35. Biol.& distrib. of Rhizopertha dominica<Tr.R.Ent.Soc.Lond.,83,449-482.

Pratt, R. Y.
    38. One hour's coll. of Scaphinotus on<Pan-Pac.Ent.,14,p.167.
    Whidby

Pratt, R. Y. and Hatch, M. H.
    38. Food of black widow spider...   <Jl.N.Y.Ent.Soc.,46,191-193.

Prebble, M. L.
    33. Larval development of 3 bark beetles   <Can.Ent.,65,145-150.

Prell, H.
    36. Beitr. zur Kennt. der Dynastinen   <Ent.Blatter,32,145-152.
    36. Beitr. zur Kennt. der Dynastinen XVI   <Deutche Ent.Zeit.,1936,179-190.

Prochazka, R.
    36. Import. morph. & syst. de la nerv. des<Sbornik Ent.Nar.Mus.Praze,1936,
    Malacoderms   14,100-132.

Procter, W.
    38. The Insect Fauna...   <Biol.Surv.Mt.Desert Region, Pt.
                     VI, Phila.,1938, 496 pp.

Putzeys, J. A. A. H.
    45. Monog. des Clivina...   <Mem.Soc.Sc.Liege,2,521-663.

Rau, G. J.
    35. N.var. of Anoplodera vittata   <Bull.Br.Ent.Soc,30,63-64.

Ray, E.
    36. Studies on N.A. Mordellidæ I   <Can.Ent.,68,124-129.

Redtenbacher, W.
    42. Quædam gen. et sp. Col...   <Vindobonæ,1842, 31 pp.

Reichardt, A.
    26. Ueber die mit Pachylobus verw. Gatt.   <Ent.Blatter,22,12-18.

Reitter, E.
    73. Syst. Eintheil. der Nitidularien   <Verh.Nat.Ver.Brunn,12,5-194.
    76. Neue exotische Nitidulidæ   <Stett.Ent.Zeit.,37,317-320.
    95. Uebersicht...Gatt. Necrophorus   <Ent.Narchrbl.,21,323-330.
    96. Best.-Tab. Eur. Col.: Carabini   <Verh.Nat.Ver.Brunn,34,36-198.
    99. Elfter Beitr. z. Col.-Fna. v. Eur.   <Wien.Ent.Zeit.,18.155-161.
    06. Cat. Col. Eur. (w. Heyden & Weise)   Paskau, 1906, 774 p.

Rey, C.
    91. Remarquez en passant   <L'Echange,7,p.19.

Riley, C. V.
  89. On Platypsyllus   (1899 instead of 1888)

Ritcher, P. O.
  36. Host relat. of Tachypterellus magnus   <Jl.Kans.Ent.Soc.,9,94-99.
  37. N.sp. of Phyllophaga from Ky.   <Ent.News,48,285-287.
  38. Field key to Ky. white grubs   <Jl.Kans.Ent.Soc.,11,24-27.
  39. Strawberry crown borer, Tyloderma   <Ky.Agr.Exp.Sta.Bull.289, 35 pp.

Ritcher, P. O., Chamberlin, T. R., and Seaton, L.
  36. Add. Rec. for Phyllophaga spreta   <Proc.Ent.Soc.Wash.,38,185-186.

Rivnay, E.
  35. Type specimen of Rhipiphorus stylopides<Ent.News,46,178-179.

Robinson, M.
  37. A new Euphoria from Texas   <Ent.News,48,p.163.
  38. Two unusual records for Penna.   <Ent.News,49,103-104.
  38. Studies in Scarabæidæ I.   <Tr.Amer.Ent.Soc.,64,107-115.

Roelofs, W.
  73. Curcul. rec. au Japan par M. Lewis   <Ann.Soc.Ent.Belg.,16,154-193.

Ross, E. S.
  37. A n.sp. of Dendrophilus from Cal.   <Pan-Pac.Ent.,13,67-68.
  37. Studies in the genus Hister   <Pan-Pac.Ent.,13,106-108.
  38. New N.A. Histeridæ   <Ent.News,49,48-51.

Roubal, J.
  03. Fundorte ein. selt. und f. Bohmen n.<Verh.zool.-bot.Ges.Wien,53,
      Käfer   380-383.

Saalas, U.
  36. Uber Flügelg. und phyl. Entw. der<Ann.Zool.Soc.zool.-bot.Fenn.
      Ceramb.   Vanamo,4,1-198.

Sabroskay, C. W.
  34. Notes on larva...Isohydnocera   <Jl.Kans.Ent.Soc.,7,65-68.

Sahlberg, J.
  71. Othismopteryx...Colydiidæ   <Notis.Sällsk.Fna.Fl.Fenn.,11,
                                  441-444.

Salle, A.
  49. Col. nouveaux de l'Amerique   <Ann.Soc.Ent.Fr.,(2),7,297-303,
                                  419-435.

Sanderson, M. W.
  36. Phyllophaga spreta in Mo.   <Jl.Kans.Ent.Soc.,9,p.30.
  37. Three n.sp. of Phyllophaga...   <Jl.Kans.Ent.Soc.,10,14-19.
  37. A n.sp. of Phyllophaga...   <Jl.Kans.Ent.Soc.,10,66-69.
  38. Elmis columbiensis a synonym...   <Jl.Kans.Ent.Soc.,11,p.146.
  38. Monog. rev. of N.A. Stenelmis   <Univ.Kans.Sci.Bull.,25,635-717.
  39. A n.g. of Scarabæidæ...   <Jl.Kans.Ent.Soc.,12,1-15.

Sanderson, M. W. and Griffith, M. E.
  35. Monstrosities in three Col.   <Jl.Kans.Ent.Soc.,8,p.25.

Satterthwait, A. F.
  33. Two n.sp. of Calendra   <Ent.News,44,210-213.
  33. Life History of...Calendra setiger   <Jl.Econ.Ent.,26,210-217.
  36. Desc. of male of Calendra dietrichi   <Ent.News,47,p.38.

Say, T.
  34. Desc. of n.sp. of N.A. insects...   <Tr.Amer.Philos.Soc.,4,409-470.

Saylor, L. W.
  33. Two new Scarabæidæ   <Can.Ent.,65,158-159.
  33. Attraction of beetles to tar   <Pan-Pac.Ent.,9,p.182.
  33. Collecting notes   <Pan-Pac.Ent.,9,p.188.
  34. Short stud. in N.A. Scarabæidæ   <Jl.Ent.& Zool.,26,49-50.
  34. Notes on Aegialia...   <Pan-Pac.,Ent.,10,74-76.
  35. Short Stud. in N.A. Scarab.III.   <Rev. de Ent.,5,33-38.
  35. New California Serica   <Jl.Ent.& Zool.,27,1-2.
  35. Studies in Amer.Scarab. II.   <Pan.-Pac.Ent.,11,35-36.
  35. A genus new to the U.S.   <Pan-Pac.Ent.,11,p.40.
  35. A Mex. sp. new to the U.S.   <Pan-Pac.Ent.,11,p.66.
  35. A new Aphodius of cadaverinus gr.   <Pan-Pac.Ent.,11,p.80.
  35. A new Cœnonycha from Nevada   <Pan-Pac.Ent.,11,p.102.

35. A n.g. & 2 n.sp. of Col. from Cal.   <Pan-Pac.Ent.,11,132-134.
36. New Calif. & Texas Scarabs   <Jl.Ent.& Zool.,28,1-4.
37. N. Scarab genera from L. & S.Cal.   <Bull.So.Cal.Ac.Sci.,36,35-37.
37. Rev. of Calif. Cyclocephala   <Jl.Ent.& Zool.,29,67-70.
37. Necessary changes...Rhizotrogid gen. <Rev.de Ent.,7,318-322.
37. A new Texas Scarab   <Can.Ent.,68,1936 (1937),p.280.
37. ...subf. Chasmatopterinæ in N.World   <Jl.Wash.Ac.Sci.,27,531-535.
38. Rev. of subf. Oncerinæ...   <Proc.Ent.Soc.Wash.,40,99-103.
38. A new Phyllophaga from Nev.   <Proc.Ent.Soc.Wash, 40,129-131.
39. Rev. of subg. Phytalus of U.S.   <Proc.U.S.Nat.Mus.,86,157-167.

Schaeffer, J. C.
71. Icones Insectorum...   Regensburg, 3 vols, 1766-1779.
77. Elementorum Entomologicorum   Ratisbonæ, 1777, App.V.

Schaeffer, C.   ℓℓ ; . ×⋯ 3ℴ⟨ᵃ ᵘⁱⁱ,
33. Notes on Hispiní and' Cassidini...   <Pan-Pac.Ent.,9,103-109.
33. Short stud. in Chrysomelidæ   <Jl.N.Y.Ent.Soc.,41,297-325.
34. Short stud. in Chrysomelidæ   <Jl.N.Y.Ent.Soc.,41,457-480.

Schaum, H. R.
44. Obs. crit. sur...Lamell. melitophiles   <Ann.Soc.Ent.Fr.,(2),2,333-426.
49. Obs. crit. sur...Lamell. melitophiles   <Ann.Soc.Ent.Fr.,(2),7,241-295.
63. Beitr. zur Kennt...Carabicinen-Gatt.   <Berl.Ent.Zeit.,7,67-92.

Schedl, K. E.
33. New Platypodidæ from C.A. & S.A.   <Rev.de Ent.,3,163-177.
35. New Scolyt. & Platypod. from C.A. &<Rev.de Ent.,5,342-359.
   S.A.

Scheerpeltz, C.
29. Monog. der Gattung Olophrum   <Verh.zool.-bot.Ges.Wien,79,257 pp.
33. Staphylinidæ VII. Suppl.I.   <Junk Col.Cat.,pars 129.
34. Staphylinidæ VIII. Suppl.II.   <Junk Col.Cat.,pars 130.

Schenkling, S.
02. Clerides nouv. du Mus. de Paris   <Bull.Mus.Paris,8,317-333.
34. Telegeusidæ, Biphyllidæ, Aculognathidæ,<Junk Col.Cat.,pars 133.
   Hemipeplidæ, Scalidiidæ
35. Ectrephidæ, Curculionidæ: Magdalinæ <Junk Cci.Cat., pars 141.

Schenkling, S. and Marshall, G. A. K.
34. Curcul.:Anthonominæ, Læmosaccinæ   <Junk Col.Cat.,pars 139.
36. Curcul.: Prionomerinæ, Aterpinæ, Ama-<Junk Col.Cat.,pars 150.
   lactinæ, Haplonychinæ, Omophorinæ
37. Curcul.: Rhadinosominæ, Trachodinæ,<Junk Col.Cat.,pars 154.
   Raymondionyminæ

Schilling, P. S.
29. Holocnemis Carab. gen. nov.   <Beitr. Ent. schles.Fn.,1,93-94.

Schönherr, C. J.
26. Curculionidum disposito meth...   Lipsiæ, 1862, 338 pp.

Schrank, J. A.
98. Fauna Boica   Nurnberg, 3 vols., 1798-1804.

Schürhoff, P. N.
37. Beitr.zur Kenntniss der Cetoniden   <Deutsche Ent.Zeit.,1937,56-80.

Schwardt, H. H.
33. Life history of lesser grain boror   <Jl.Kans.Ent.Soc.,6,61-66.

Scopoli, J. A.
68. Annus historico naturalis   Lipsiæ, 1, 1768, 168 pp.

Scott, F. T.
33. Addit. to Coccinellidæ of Alaska   <Pan-Pac.Ent.,9,p.126.

Scott, H.
33. Syst. pos. of Hemipeplidæ...   <Ann.Mag.Nat.Hist.,(10),12,595-611.

Scriba, W.
55. Neue Staphylinen   <Stett.Ent.Zeit.,16,295-302.

Seevers, C. H.
38. Termitophilous Col. in U.S.   <Ann.Ent.Soc.Amer.,31,422-441.

Segal, B.
33. Hind wings of some Dryopidæ   <Ent.News,44,85-88.

Seidlitz, G.
  91. Fauna Baltica...   Königsberg, (Ed.2),1891,818 pp.

Semenov-Tian-Shanskij, A. P.
  26. Analecta coleopterologica   <Rev.Russ.d'Ent.,20,33-55.
  32. De tribu Necrophorini classif...   <Trav.Inst.Zool.Ac.Sci.U.R.S.S., 1,149-191
  33. Sur la distr. geog. de Nomius pygmæus<Bull.Soc.Ent.Fr.,38,194-195.

Sharp, D.
  70. Elateridæ of New Zealand   <Ann.Mag.Nat.Hist.,(4),19,396-413, 469-487.
  74. Staphylinidæ of Japan   <Tr.Ent.Soc.Lond.,1874,1-103.
  80. Col. from Hawaiian Is.   <Tr.Ent.Soc.Lond.,1880,37-54.

Sloop, K. D.
  35. Three n.sp. of Plastocerinæ...   <Pan-Pac.Ent.,11,17-20.
  35. Notes on two rare Elateroids   <Pan-Pac.Ent.,11,p.24.
  35. Distrib. notes on Cal. Elateridæ <Pan-Pac.Ent.,11,p.64.
  37. Rev. of N.A...Melanophila   <Univ.Cal.Pub.Ento.,7,1-20.

Smith, O. J.
  31. Study of Tenebrrionidæ of s.e.Iowa <Proc.Iowa Ac.Sci.,38,259-265.

Smyth, E. G.
  33. On Nic. & Weiss Syn. of Cicindelidæ <Ent.News,44,197-204.
  34. Gregarious habit in beetles   <Jl.Kans.Ent.Soc.,7,102-119.
  35. Analysis of Cicindela purpurea gr. <Ent.News,46,14-19, 44-49.

Snyder, T. E.
  35. Introduced Anobiid destructive... <Proc.Biol.Soc.Wash.,48,59-60.

Solsky, S.
  68. Etudes sur Staphyl. du Mexique   <Hor.Soc.Ent.Ross.,5,119-144.

Spaeth, F.
  36. Mitteil. uber...Cassidinen...   <Ent.Rundschau,53,65-69, 138-140, 170-173, 213-216, 259-262.
  37. Mitteil. uber Cassidinen...Stettin <Stett.Ent.Zeit.,98,79-96.

Spinola, M. M.
  41. Monog. des Terediles...   <Rev.Zool.,4,70-76.

Stirrett, G. M.
  33. N. flea-beetle from Iowa   <Can.Ent,65,208-210.

Strand, E.
  28. Nomencl. Bemerk...Col.-Gatt.,   <Ent.Nachrbl.,2,2-3.
  36. Misc. nomencl. zool. et palæont. <Folia zool. Hydrobiol.,9,167-170.

Suffrian, E.
  40. Fragm...deutscher Käfer   <Stett.Ent.Zeit.,1,82-86,98-104.
  41. Fragm...deutscher Käfer   <Stett.Ent.Zeit.,2,19-25,38-47,etc.

Sulzer, J. H.
  61. Die Kennzeichen der Insekten... Zürich, 1761, 204 + 67 pp.

Swaine, J. M.
  34. Three n.sp. of Scolytidæ   <Can.Ent.,66,204-206.

Szekessy, V.
  36. Uber...ocellen der...Pteroloma   <Ann.Mus.Nat.Hung.,30,Zool.,48-49.

Tanner, V. M.
  34. Col. of Zion Nat. Park, No.II.   <Ann.Ent.Soc.Amer.,27,43-49.
  34. Stud. in weevils of w.U.S. I.   <Proc.Utah Ac.Sci.,11,285-288.
  35. List of ins. types at B.Y.U.,Provo, No.I.<Proc.Utah Ac.Sci.,12,181-193.
  36. Desc. of two n. Melyrids from Utah <Proc.Utah Ac.Sci.,13,153-154.
  36. List of ins.types...II.   <Proc.Utah Ac.Sci.,13.147-152.
  38. A n. weevil...Dyslobus, II.   <Proc.Utah Ac.Sci.,15,147-148.

Tanner, V. M. and Hayward, C. L.
  34. Biol. study of LaSal Mts., Utah <Proc.Utah Ac.Sci.,11,209-233.

Thatcher, T. O.
  35. Scolytidæ of Logan Canyon area <Proc.Utah Sc.Sci.,12,261-262.

Thomas, C. A.
  33. Prothetely in an Elaterid larva <Ent.News,44.91-96.

Thunberg, G. A.
  15. De Coleopteris rostratis   <Nova Acta Upsal.,7,104-125.

Ting, P. C.
33. Feeding mechanisms of weevils   <Mo.Bull.Cal.Dpt.Agr.,22,161-165.
34. (Potosia afflnis)...at San Francisco   <Mo.Bull.Cal.Dpt.Agr.,23,185-191.
36. Pupation of Haltica bimarginata   <Pan.Pac.Ent.,12,p.55.
36. Mouthparts of...Rhynchophora   <Microent.,1,93-114.
37. Collecting notes   <Pan-Pac.Ent.,13,p.24.
37. A n.sp. of Dyslobus...   <Bull.So.Cal.Ac.Sci.,36,79-83.
38. A n.sp. of Panscopus...   <Pan-Pac.Ent.,14,121-123.

Travis, B. V.
34. Phyllophaga of Iowa   <Iowa St.Coll.Jl.Sci.,8,313-365.

Trippel, A. W.
34. New rec. of Ind. Chrysomelidæ   <Bull.Br.Ent.Soc.,29,74-76.

Twinn, C. R.
34. Dermestid Trogoderma versicolor   <Can.Ent.,66,49-51.

Uhmann, E.
36. Amer. Hispinen: Chalepus   <Festschr.E.Strand,1,611-629.
38. Amer. Hispinen: Xenochalepus   <Rev.de Ent.,8,420-440.

Valentine, J. M.
34. Technique in prepar. of Col.   <Jl.E.Mitchell Sci.Soc.,50,255-262.
35. Speciation in Steniridia...   <Jl.E.Mitchell Sci.Soc.,51,341-375.
36. Raciation in Steniridia andrewsi...   <Jl.E.Mitchell Sci.Soc.,52,223-234.
37. Anophthalmid Col. from Tenn.   <Jl.E.Mitchell Sci.Soc.,53,93-100.

Van Dyke, E. C.
26. Value of hife hist. stud. in tax.   <Jl.Econ.Ent.,19,703-707.
27. Secondary sex. char. in Col.   <Proc.Pac.Cst.Ent.Soc.,2,75-84.
32. Col. fna. of semiarid s.w. N.A.   <V Int.Cong.Ent.,1932 (1933),
          471-477.
33. Rev. of Dyslobus   <Pan-Pac.Ent.,9,31-47.
33. Two n.sp. of Scarabæidæ   <Pan-Pac.Ent.,9,115-116.
33. A n.sp. of Pleocoma   <Pan-Pac.Ent.,9,183-184.
34. N.spp. of Buprestidæ   <Ent.News,45,61-66,89-91.
34. A root-boring Derobrachus   <Pan-Pac.Ent.,10,p.58.
34. PP. 323-335 in Termites & termite<Rpt.Term.Invest.Comm.,
    control   Berkeley,1934.
34. Note on Liebeck collection   <Pan-Pac.Ent.,10,p.158.
34. N.sp. of N.A....Brachyrhininæ   <Pan-Pac.Ent.,10,175-191.
34. N.A. spp. of Trigonurus   <Bull.Br.Ent.Soc.,29.177-182.
35. N.spp. of N.A. Brachyrhininæ,II.   <Pan-Pac.Ent.,11,1-10.
35. N.spp. of N.A. Brachyrhininæ,III <Pan-Pac.Ent.,11,83-96.
36. N.spp. of N.A. Brachyrhininæ,IV. <Pan-Pac.Ent.,12,19-32.
36. N.spp. of N.A. Brachyrhininæ,V   <Pan-Pac.Ent.,12.73-85.
36. Pp. 160-255, etc. in Forest Insects.   New York, 1936.
36. Rev. of subg. Nomaretus...   <Bull.Br.Ent.Soc.,31,37-44.
36. Another destr. deathwatch beetle   <Pan-Pac.Ent.,12,p.178.
36. A correction   <Pan-Pac.Ent.,12,p.183.
36. Change of name   <Pan-Pac.Ent.,12,p.191.
37. Notes& desc. of N.A. Buprest.& Ceramb. <Bull.Br.Ent.Soc.,32.105-116.
37. Weevil larvæ annoying householders   <Pan-Pac.Ent.,13,p.93.
37. Eudiagogus pulcher   <Pan-Pac.Ent.,13,p.170.
38. N.spp. of Rhynchophora of w.N.A.   <Pan-Pac.Ent.,14-1-9.
38. Rev. of Chrysolina...   <Bull.Br.Ent.Soc.,33,45-58.
38. Carabus forreri in Arizona   <Pan-Pac.Ent.,14,p.95.
38. Rev. of subg. Scaphinotus...   <Ent.Amer.,18,93-133.
38. Calendra (Sphenophorus) minimus in<Pan-Pac.Ent.,14,p.187.
    Cal.
38. N.spp. of Pac. Cst. Coleoptera   <Ent.News,49,189-195.

Van Emden, F.
36. Klassif. der...Carab. & Harpal. piliferæ<Ent.Blatter,32.12-17.
38. Taxon. of Rhynchophora larvæ   <Tr.R.Ent.Soc.Lond.,87,1-37.

Van Emden, M. and Van Emden, F.
—. Curcul.: Brachyderinæ III.   <Junk Col.Cat.,pars 164, in press.

Voris, R.
34. Biol. investig. on Staphylinidæ   <Tr.Ac.Sci.St.Louis,28.233-261.
36. Rapid spread of Eur. Staph. in N.A. <Ann.Ent.Soc.Amer.,29,78-80.

Voss, E.
22. Indo-Malayische Rhynchitinen   <Phil.Jl.Sci.,21,385-413.
33. Monog. der...Auletini...   <Stett.Ent.Zeit.,94,108-136,273-286.
34. Einige unbeschr. neotrop. Curcul...   <Sbornik ent.Nar.Mus.Praze,12, 63-104.
34. Monog. der...Auletini. .   <Stett.Ent.Zeit.,95,109-135,330-344.
35. Monog. der...Auletini...   <Stett.Ent.Zeit.,96,91-105,229-241.
36. Monog. der...Auletini...   <Stett.Ent.Zeit.,97,279-289.
37. Monog. der...Auletini...   <Stett.Ent.Zeit.,98,101-108.
38. Monog. der...Deporaini...   <Stett.Ent.Zeit.,99,59-117.
38. Monog. der...Rhynchitini...   <Kol.Rundschau,24,129-171.
—. Curcul.: Rhynchitinæ II.   <Junk Col.Cat., in press.

Wade, J. S.
33. Beetles that stand on their heads   <Nat.Mag.,22,213-215.
35. Contr. to bibliog. of immat. N.A. Col.   Washington, 1935, mimeo.,114 pp.

Walker, F.
59. Char. of undescr. Ceylon ins.   <Ann.Mag.Nat.Hist.,(3),3,50-56, 256-265.

66. Pp. 312-334 (see original cat. biblio.)

Wallis, J. B.
32. Rev. N.A...Haliplus   <Tr.Roy.Canad.Inst.,19,1-76.
33. N.sp. of Hypophlœus   <Can.Ent.,65,247-249.
33. Three n.sp. of Hydroporus...   <Can.Ent.,65,261-262.
33. Some new Dytiscidæ   <Can.Ent.,65,268-278.

Wankowicz, J.
67. Notices sur divers Coléoptères   <Ann.Soc.Ent.Fr.,(4),7,249-255.

Wasmann, E.
03. My last reply to Major Casey   <Can.Ent.,35,74-75.

Waterhouse, C. O.
75. On Lamellicorn Col. of Japan   <Tr.Ent.Soc.Lond.,1875,71-116.
76. N.spp. of Col. from Rodriguez...   <Ann.Mag.Nat.Hist.,(4),18,105-121.
77. Desc. of n.Col. from various loc.   <Ent.Mo.Mag.,(1),14,23-28.
95. Desc. of n.Col. in the Br. Museum   <Ann.Mag.Nat.Hist.,(6),16,157-160.

Watson, J. R.
37. Naupactus leucoloma...in U.S.   <Florida Ent.,20-1-3.

Weise, J.
95. Neue Coccinelliden...   <Ann.Soc.Ent.Belg.,39,120-146.

Weiss, H. B.
14. Agrilus politus infesting roses   <Jl.Econ.Ent.,7,438-440.

Wenzel, H. W.
96. Notes on Lampyridæ   <Ent.News,7,294-296.

Wenzel, R. L.
35. Sexual characters of Saprinus...   <Can.Ent.,67,189-190.
37. Short studies in Histeridæ I.   <Can.Ent.,68,1936 (1937),266-272.
39. Short studies in Histeridæ 2.   <Chio Jl.Sci.,39,10-14.

Westwood, J. O.
41. Insectorum novorum Centuria   <Ann.Mag.Nat.Hist.,8.203-205.
45. On Lamell. beetles which possess...   <Tr.Ent.Soc.Lond.,4,155-180.
49. in White's Cat. Clerid, 1849, p. 50.
75. On spp. of Rutelidæ...E.Asia & Malay <Tr.Ent.Soc.Lond.,1875,233-239.

Whelan, D. B.
36. A flea beetle new to Nebraska   <Jl.Kans.Ent.Soc.,9,p.30.
36. Col. of original prairie in e.Neb.   <Jl.Kans.Ent.Soc.,9,111-115.

White, A.
46. Ins. in Richardson & Gray, Zool. of Voyage of Erebus & Terror   London, 1846, pp. 1-24.

White, B. E.
37. Notes on...Agabus lineelus   <Pan-Pac.Ent.,13,p.84.
37. Three new...Cryptocephalus   <Pan-Pac.Ent.,13,111-114.

Wickham, H. F.
10. List of Van Duzee...Fla. beetles   <Bull.Buff.Soc.N.S.,9. 399-405.
11. List of Col. of Iowa   <Bull.Lab.Iowa,6,no.2,1-40.

Wiedemann, C. R. W. and Germar, E. F.
  21. Neue exotische Kafer   <Mag.Ent.,4,107-183.
Wilcox, J. and Baker, W. W.
  35. Deciduous cusps of Alophini   <Bull.Br.Ent.Soc.,30,20-21.
Williams, I. W.
  38. Comp. morph. of mouthparts of Col. <Jl.N.Y.Ent.Soc.,46,245-289.
Wilson, S. J.
  34. Anat. of Chrysochus auratus...   <Jl.N.Y.Ent.Soc.,42,65-85.
Wolcott, A. B. and Montgomery, B. E.
  33. Ecol. study of Col. of Tamarack swamp<Amer.Midl.Nat.,14,113-169.
Wolcott, G. N.
  37. Animal census of pastures & meadow <Ecol.Monogr.,7,1-90.
Wolfrum, P.
  38. Beitrg. zur Kennt. der Anthribiden   <Ent.Blatter,34,67-76.
Zaitzev, P. A.
  06. Notizen über Wasserkäfer...   <Rev.Russ.d'Ent.,6,170-175.
Zetterstedt, J. W.
  24. Nagra nya Svenska Insect-arter... <Vetensk.Acad.Handl.,1924,149-159.
Zia, Y.
  37. Comp. stud. of male gen. tube in Col.<Sinensia,7,319-352.
Zimmerman, E. C.
  36. Brachytarsus in California   <Pan-Pac.Ent.,12,p.191.

THIRD SUPPLEMENT TO
CATALOGUE OF NORTH AMERICAN COLEOPTERA
DESCRIBED AS FOSSILS

The few fossil beetles described as new in the years 1933 to 1938 inclusive are listed below. Names in brackets were invalid as published. One new abbreviation is used.

Ariz. Trias.   Chinle Formation, Petrified Forest National Monument, Holbrook, Arizona.

CARABIDÆ

Scaphinotus Dej.
   minor Horn 76-243, belongs in subg. Irichroa [1].
[1] Lapouge—32.
   wheatleyi Horn 76-242, belongs in subg. Scaphinotus [1].

Carabus Linn.
   mæander Fisch., belongs in sub. Eucarabus [2].
   v.sangamon Wickh., belongs in subg. Eucarabus [2].
[2] Breuning, 32-37.

Patrobus Dej. [3]
[3] Revision of genus, Darlington—38.

gelatus Scudd. 90-530 [3]	Tor.[4] Scar. Pleist.
decessus Scudd. 00-73 [3]	Scar. Tor.[4] Pleist.
frigidus Scudd. 00-74 [3]	Tor.[4] Pleist.
†*stygicus* Chd. 71-46 [3]	Recent.
henshawi Wickh. 17-140	Ill. Pleist.

[4] Coleman—33. as valid species.

Elaphrus Fab.	
irregularis Scudd.	Tor.[4] Scar. Pleist.
Loricera Latr.	
glacialis Scudd.	Tor.[4] Scar. Pleist.
lutosa Scudd.	Tor.[4] Scar. Pleist.
exita Scudd.	Tor.[4] Scar. Pleist.
Bembidium Latr.	
glaciatum Scudd.	Tor.[4] Scar. Pleist.
vestigium Scudd.	Tor.[4] Scar. Pleist.
vanum Scudd.	Tor.[4] Scar. Pleist.
præteritum Scudd.	Tor.[4] Scar. Pleist.
damnosum Scudd.	Tor.[4] Scar. Pleist.
Pterostichus Bon.	
abrogatus Scudd.	Tor.[4] Scar. Pleist.
destitutus Scudd.	Tor.[4] Scar. Pleist.
fractus Scudd.	Tor.[4] Scar. Pielst.
destructus Scudd.	Tor.[4] Scar. Pleist.
gelidus Scudd.	Tor.[4] Scar. Pleist.
depletus Scudd.	Tor.[4] Scar. Pleist.
Platynus Bon.	
casus Scudd.	Tor.[4] Scar. Pleist.
hindei Scudd.	Tor.[4] Scar. Pleist.
halli Scudd.	Tor.[4] Scar. Pleist.
dissipatus Scudd.	Tor.[4] Scar. Pleist.
desuetus Scudd.	Tor.[4] Scar. Pleist.
harttii Scudd.	Tor.[4] Scar. Pleist.
delapidatus Scudd.	Tor.[4] Scar. Pielst.
exterminatus Scudd.	Tor.[4] Scar. Pleist.
interglacialis Scudd.	Tor.[4] Scar. Pleist.
longævus Scudd.	Tor.[4] Scar. Pleist.
Harpalus Latr.	
conditus Scudd.	Tor.[4] Scar. Pleist.

DYTISCIDÆ

Cœlambus Thoms.
  derelictus Scudd.   Tor.[4] Scar. Pleist.
  disjectus Scudd.   Tor.[4] Scar. Pleist.

Hydroporus Clairv.
  inanimatus Scudd.   Tor.[4] Scar. Pleist.
  sectus Scudd.   Tor.[4] Scar. Pleist.

Agabus Leach
  perditus Scudd.   Tor.[4] Scar. Fielst.

GYRINIDÆ

Protogyrininus Hatch
  confinis Lec.   Tor. Pleist.[4]

HYDROPHILIDÆ

Cymbiodyta Bedel
  extincta Scudd.   Tor.[4] Scar. Pleist.

STAPHYLINIDÆ

Olophrum Er.
  celatum Scudd.   Tor.[4] Scar. Pleist.
  arcanum Scudd.   Tor.[4] Scar. Pleist.
  dejectum Scudd.   Tor.[4] Scar. Pleist.

Arpedium Er.
  stillicidii Scudd.   Tor.[4] Scar. Pleist.

Acidota Mannh.
  crenata Fabr.
    v.nigra Scudd.   Tor.[4] Scar. Pleist.

Geodromicus Redt.
  stiricidii Scudd.   Tor.[4] Scar. Pleist.

Bledius Mannh.
  glaciatus Scudd.   Tor.[4] Scar. Fielst.

Oxyporus Fabr.
  stiriacus Scudd.   Tor.[4] Scar. Pleist.

Lathrobium Grav.
  frustrum Scudd.   Tor.[4] Scar. Pleist.
  inhibitum Scudd.   Tor.[4] Scar. Pleist.
  debilitatum Scudd.   Tor.[4] Scar. Pleist.
  interglaciale Scudd.   Tor.[4] Scar. Pleist.

Cryptobium Mannh.
  cinctum Scudd.   Tor.[4] Scar. Fielst.
  detectum Scudd.   Tor.[4] Scar. Pleist.

Philonthus Curt.
  claudus Scudd.   Tor.[4] Scar. Pleist.

Quedius Steph.
  deperditus Scudd.   Tor.[4] Scar. Pleist.

BUPRESTIDÆ

[Paleobuprestis Walker 38-138 [5]]
   [maxima Walker 37-138 [5]   Ariz. Trias.]
   [minima Walker 38-139 [5]   Ariz. Trias.]
[5] Walker—38; these names are invalid because they are not based upon specimens of insects but rather upon reputed evidence that the insects once existed.

CHRYSOMELIDÆ

Donacia Fab.
   stiria Scudd.   Tor.[4] Scar. Pleist.
   pompatica Scudd.   Tor.[4] Scar. Pleist.

SCOLYTIDÆ

[Paleoscolytus Walker 38-139 [5]]
   [divergus Walker 38-139[5]   Ariz. Trias.]

[Paleoipidus Walker 38-140 [5]]
   [perforatus Walker 38-140 [5]   Ariz. Trias.]
   [marginatus Walker 38-140 [5]   Ariz. Trias.]

BIBLIOGRAPHY

Coleman
   33. Pleist. of Toronto Region.   <41st Ann.Rpt.Ont.Dept. Mines. pt.7,69 pp.

Darlington
   38. American Patrobiini.   <Ent.Amer.,18,135-187.

Walker
   38. Triassic ins. in Arizona.   <Proc.U.S.N.M.,85,137-141.

INDEX

In order to bring together all the references to each name in the catalog and the supplements, this index is a combination of all the previous indexes and the one for the new supplement. It can be used in place of the other indexes because it is more complete than each. The pages in the supplements are distinguished by superscript numerals thus: 24 — original catalogue, 24[1] — first supplement, 24[2] — second supplement, 96[3] — third supplement and 24[4] — fourth supplement. (The second and third supplements were paged consecutively.) *Romans ..... .a i refer to page ... .; .. .. .: .main catis 3ue.*